T0230446

Statistics and Computing

Series editor

W.K. Härdle, Humboldt-Universität zu Berlin, Berlin, Germany

Statistics and Computing (SC) includes monographs and advanced texts on statistical computing and statistical packages.

More information about this series at http://www.springer.com/series/3022

Wolfgang Karl Härdle · Ostap Okhrin
Yarema Okhrin

Basic Elements
of Computational Statistics

 Springer

Wolfgang Karl Härdle
CASE – Center for Applied Statistics
and Economics, School of Business
and Economics
Humboldt-Universität zu Berlin
Berlin
Germany

Yarema Okhrin
Chair of Statistics, Faculty of Business
and Economics
University of Augsburg
Augsburg
Germany

Ostap Okhrin
Econometrics and Statistics, esp. in
Transportation, Institut für Wirtschaft und
Verkehr, Fakultät Verkehrswissenschaften
"Friedrich List"
Technische Universität Dresden
Dresden, Sachsen
Germany

ISSN 1431-8784   ISSN 2197-1706   (electronic)
Statistics and Computing
ISBN 978-3-319-85631-5   ISBN 978-3-319-55336-8   (eBook)
DOI 10.1007/978-3-319-55336-8

Mathematics Subject Classification (2010): 62-XX, 62G07, 62G08, 62H15, 62Jxx

Printed on acid-free paper

This Springer imprint is published by Springer Nature
The registered company is Springer International Publishing AG
The registered company address is: Gewerbestrasse 11, 6330 Cham, Switzerland

To Martin, Sophie, and Iryna, who are my grace and inspiration

Ostap

To Katharine, Justine, Danylo, and Irena for making my life colorful

Yarema

To our parents

Ostap and Yarema

To my family

Wolfgang

Учітесь, читайте, і чужому научайтесь, й свого не цурайтесь.

–Тарас Шевченко

Think and read, and to your neighbours' gifts
pay heed, yet do not thus neglect your own.

—Taras Shevchenko
(translated by C. H. Andrusyshen and W. Kirkconnel)

Preface

The R programming language is becoming the lingua franca of computational statistics. It is the usual statistical software platform used by statisticians, economists, engineers and scientists both in corporations and in academia. Established international companies use R in their data analysis. R has gained its popularity for two reasons. First, it is an OS independent free open-source program which is popularised and improved by hundreds of volunteers all over the world. A plethora of packages are available for many scientific disciplines. Second, common analysts can do complicated analyses without deep computer programming knowledge. This book on the basic elements of computational statistics presents the tools and concepts of univariate and multivariate data analyses with a strong focus on applications and implementations. The aim of this book is to present data analysis in a way that is understandable for non-mathematicians and practitioners who are confronted by statistical data analysis. All practical examples may be recalculated and modified by the reader: all data sets and programmes (Quantlets) used in the book are downloadable from the publisher's home page of this book (www.quantlet.de). The text contains a wide variety of exercises and covers the basic mathematical, statistical and programming problems.

The first chapter introduces the reader to the basics of the R language, taking into account that only minimal prior experience in programming is required. Starting with the developing history and R environments under different operating systems, the book discusses the syntax. We start the description of the syntax with the classical 'Hello World!!!' program. The use of R as an advanced calculator, data types, loops, if then conditions, own function construction and classes are the topics covered in this chapter. As in statistical analysis one deals with data, special attention is paid to work with vectors and matrices.

The second part deals with the numerical techniques which one needs during the analysis. A short excursion into matrix algebra will be helpful in understanding multivariate techniques provided in the further sections. Different methods of numerical integration, differentiation and root finding help the reader to get inside the core of the R system.

Chapter 3 highlights set theory, combinatoric rules, plus some of the main discrete distributions: binomial, multinomial, hypergeometric and Poisson.

Different characteristics, cumulative distribution functions and density functions of the continuous distributions: uniform, normal, t, χ^2, F, exponential and Cauchy will be explained in detail in Chapter 4.

The next chapter is devoted to univariate statistical analysis and basic smoothing techniques. The histogram, kernel density estimator, graphical representation of the data, confidence intervals, different simple tests as well as tests that need more computations, like the Wilcoxon, Kruskal–Wallis, sign tests, are the topics of Chapter 5.

The sixth chapter deals with multivariate distributions: their definition, characteristics and application of general multivariate distributions, multinormal distributions, as well as classes of copulas. Further, Chapter 7 discusses linear and nonlinear relationships via regression models.

Chapter 8 partially extends the problems solved in Chapter 5, but also considers more sophisticated topics, such as multidimensional scaling, principal component, factor, discriminant and cluster analysis. These techniques are difficult to apply without computational power, so they are of special interest in this book.

Theoretical models need to be calibrated in practice. If there is no data available, then Monte Carlo simulation techniques are necessary parts of each study. Chapter 9 starts from simple sampling techniques from the uniform distribution. These are further extended to simulation methods from other univariate distributions. We also discuss simulation from multivariate distributions, especially copulae.

Chapter 10 describes more advanced graphical techniques, with special attention to three-dimensional graphics and interactive programmes using packages lattice, rgl and rpanel.

This book is designed for the advanced undergraduate and first-year graduate student as well as for the inexperienced data analyst who would like a tour of the various statistical tools in a data analysis workshop. The experienced reader with a good knowledge of statistics and programming will certainly skip some sections of the univariate models, but hopefully enjoy the various mathematical roots of the multivariate techniques. A graduate student might think that the first section on description techniques is well known to him from his training in introductory statistics. The programming, mathematical and the applied parts of the book will certainly introduce him into the rich realm of statistical data analysis modules.

A book of this kind would not have been possible without the help of many friends, colleagues and students. For many suggestions, corrections and technical support, we would like to thank Aymeric Bouley, Xiaofeng Cao, Johanna Simone Eckel, Philipp Gschöpf, Gunawan Gunawan, Johannes Haupt, Uri Yakobi Keller, Polina Marchenko, Félix Revert, Alexander Ristig, Benjamin Samulowski, Martin Schelisch, Christoph Schult, Noa Tamir, Anastasija Tetereva, Tatjana Tissen-Diabaté, Ivan Vasylchenko and Yafei Xu. We thank Alice Blanck and Veronika Rosteck from Springer Verlag for continuous support and valuable suggestions on the style of writing and the content covered. Special thanks go to the anonymous proofreaders who checked not only the language but also the statistical, programming and mathematical content of the book. All errors are our own.

Berlin, Germany   Wolfgang Karl Härdle
Dresden, Germany   Ostap Okhrin
Augsburg, Germany   Yarema Okhrin
April 2017

Contents

Symbols and Notations

Basics

X, Y	Random variables or vectors
$X_1, X_2, \ldots, X_p$	Random variables
$X = (X_1, \ldots, X_p)^{\top}$	Random vector
$X \sim \cdot$	X has distribution $\cdot$
Γ, Δ	Matrices
$\mathcal{A}, \mathcal{B}, \mathcal{X}, \mathcal{Y}$	Data matrices
Σ	Covariance matrix
1_n	Vector of ones $(\underbrace{1, \ldots, 1}_{n-times})^{\top}$
0_n	Vector of zeros $(\underbrace{0, \ldots, 0}_{n-times})^{\top}$
$\mathcal{I}_n$	Identity matrix
$\mathrm{I}(.)$	Indicator function
$\lceil \ldots \rceil$	Ceiling function
$\lfloor \ldots \rfloor$	Floor function
i	Imaginary unit, $i^2 = -1$
$\Rightarrow$	Implication
$\Leftrightarrow$	Equivalence
$\approx$	Approximately equal
iff	if and only if, equivalence
i.i.d.	Independent and identically distributed
rv	Random variable
$\mathbb{R}^n$	n-dimensional space of real numbers

δ_{ik}	The Kronecker delta, that is 1 if $i = k$ and 0 otherwise
P_n	$P_n = \{v \in C[a,b] \| v(x) = \sum_{i=0}^{n} a_i x^i, a_i \in \mathbb{R}\}$
$f(x) \in \mathcal{O}\{g(x)\}$	There is $k > 0$ such that for all sufficiently large values of x, $f(x)$ is at ost $kg(x)$ in absolute value
$\mathrm{med}\,(x)$	The median value of the sample x

Samples

x, y	Observations of X and Y
$x_1, \ldots, x_n = \{x_i\}_{i=1}^{n}$	Sample of n observations of X
$\mathcal{X} = \{x_{ij}\}_{i=1,\ldots,n; j=1,\ldots,p}$	$(n \times p)$ data matrix of observations of $X_1, \ldots, X_p$ or of $X = (X_1, \ldots, X_p)^{\top}$
$x_{(1)}, \ldots, x_{(n)}$	The order statistic of $x_1, \ldots, x_n$
$\mathcal{H}$	Centering matrix, $\mathcal{H} = \mathcal{I}_n - n^{-1} 1_n 1_n^{\mathrm{T}}$
$\bar{x}$	The sample mean

Densities and Distribution Functions

$f(x)$	Density of X
$f(x, y)$	Joint density of X and Y
$f_X(x), f_Y(y)$	Marginal densities of X and Y
$f_{X_1}(x_1), \ldots, f_{X_p}(x_p)$	Marginal densities of $X_1, \ldots, X_p$
$\hat{f}_h(x)$	Histogram or kernel estimator of $f(x)$
$F(x)$	Distribution function of X
$F(x, y)$	Joint distribution function of X and Y
$F_X(x), F_Y(y)$	Marginal distribution functions of X and Y
$F_{X_1}(x_1), \ldots, F_{X_d}(x_d)$	Marginal distribution functions of $X_1, \ldots, X_d$
$\phi_X(t)$	Characteristic function of X
m_k	k-th moment of X
$\hat{F}(x)$	Empirical cumulative distribution function (ecdf)
pdf	Probability density function

Empirical Moments

$\bar{x} = \dfrac{1}{n} \sum_{i=1}^{n} x_i$	Average of X sampled by $\{x_i\}_{i=1,\ldots,n}$
$s_{XY}^2 = \dfrac{1}{n-1} \sum_{i=1}^{n} (x_i - \bar{x})(y_i - \bar{y})$	Empirical covariance of random variables X and Y sampled by $\{x_i\}_{i=1,\ldots,n}$ and $\{y_i\}_{i=1,\ldots,n}$
$s_{XX}^2 = \dfrac{1}{n-1} \sum_{i=1}^{n} (x_i - \bar{x})^2$	Empirical variance of random variable X sampled by $\{x_i\}_{i=1,\ldots,n}$
$r_{XY} = \dfrac{s_{XY}^2}{\sqrt{s_{XX}^2 s_{YY}^2}}$	Empirical correlation of X and Y

$\hat{\Sigma} = \{s_{X_iX_j}\}$ — Empirical covariance matrix of a sample or observations of $X_1,\ldots,X_p$ or of the random vector $X = (X_1,\ldots,X_p)^\top$

$\mathcal{R} = \{r_{X_iX_j}\}$ — Empirical correlation matrix of a sample or observations of $X_1,\ldots,X_p$ or of the random vector $X = (X_1,\ldots,X_p)^\top$

Distributions

$\varphi(x)$ — Density of the standard normal distribution

$\Phi(x)$ — Cumulative distribution function of the standard normal distribution

$N(0,1)$ — Standard normal or Gaussian distribution

$N(\mu,\sigma^2)$ — Normal distribution with mean μ and variance σ^2

$N_d(\mu,\Sigma)$ — d-dimensional normal distribution with mean μ and covariance matrix Σ

$\xrightarrow{\mathcal{L}}$ — Convergence in distribution

$\xrightarrow{a.s}$ — Almost sure convergence

$\overset{a}{\sim}$ — Asymptotic distribution

$U(a,b)$ — Uniform distribution on (a,b)

CLT — Central Limit Theorem

χ^2_p — χ^2 distribution with p degrees of freedom

$\chi^2_{1-\alpha;p}$ — $1-\alpha$ quantile of the χ^2 distribution with p degrees of freedom

t_n — t-distribution with n degrees of freedom

$t_{1-\alpha/2;n}$ — $1-\alpha/2$ quantile of the t-distribution with n d.f

$F_{n,m}$ — F-distribution with n and m degrees of freedom

$F_{1-\alpha;n,m}$ — $1-\alpha$ quantile of the F-distribution with n and m degrees of freedom

$B(n,p)$ — Binomial distribution

$H(x;n,M,N)$ — Hypergeometric distribution

$\text{Pois}(\lambda_i)$ — Poisson distribution with parameter λ_i

Mathematical Abbreviations

$\text{tr}(\mathcal{A})$ — Trace of matrix $\mathcal{A}$

$\text{diag}(\mathcal{A})$ — Diagonal of matrix $\mathcal{A}$

$\text{rank}(\mathcal{A})$ — Rank of matrix $\mathcal{A}$

$\det(\mathcal{A})$ — Determinant of matrix $\mathcal{A}$

id — Identity function on a vector space V

$C[a,b]$ — The set of all continuous differentiable functions on the interval $[a,b]$

Chapter 1
The Basics of R

Don't think—use the computer.

— G. Dyke

1.1 Introduction

The R software package is a powerful and flexible tool for statistical analysis which is used by practitioners and researchers alike. A basic understanding of R allows applying a wide variety of statistical methods to actual data and presenting the results clearly and understandably. This chapter provides help in setting up the programme and gives a brief introduction to its basics.

R is open-source software with a list of available, add-on packages that provide additional functionalities. This chapter begins with detailed instructions on how to install it on the computer and explains all the procedures needed to customise it to the user's needs.

In the next step, it will guide you through the use of the basic commands and the structure of the R language. The goal is to give an idea of the syntax so as to be able to perform simple calculations as well as structure data and gain an understanding of the data types. Lastly, the chapter discusses methods of reading data and saving datasets and results.

1.2 R on Your Computer

1.2.1 History of the R Language

R is a complete programming language and software environment for statistical computing and graphical representation. R is closely related to S, the statistical program-

© Springer International Publishing AG 2017
W.K. Härdle et al., *Basic Elements of Computational Statistics*,
Statistics and Computing, DOI 10.1007/978-3-319-55336-8_1

ming language of Bell Laboratories developed by Becker and Chamber in 1984. It is actually an implementation of S with lexical scoping semantics inspired by Scheme, which started in 1992 and with the first results published by the developers Ihaka and Gentleman (1996) of the University of Auckland, NZ, for teaching purposes. Its name, R, is taken from the first names of the authors.

As part of the GNU Project, the source code of R has been freely available under the GNU General Public License since 1995. This decision contributed to spreading the software within the community of statisticians using free-code operating systems (OS). It is now a multi-platform statistical package widely known by people from many scientific fields such as mathematics, medicine and biology.

R enables its users to handle and store data, perform calculations on many types of variables, statistically analyse information under different aspects, create graphics and execute programmes. Its functionalities can be expanded by importing packages and including code written in C, C++ or Fortran. It is freely available on the Internet using the CRAN mirrors (Comprehensive R Archive Network at http://cran.r-project. org/). Since this chapter deals with installation issues and the basics of the R language, the reader familiar with the basics may skip it.

There exist several books about R, discussing specific topics in statistics and econometrics (biostatistics, etc.) or comparing R with other software, for example Stata. Typical users of Stata may be interested in Muenchen and Hilbe (2010). If the research topic requires Bayesian econometrics and MCMC techniques, Albert (2009) might be helpful. Two additional books on R, by Gaetan and Guyon (2009) and Cowpertwait and Metcalfe (2009), may support the development of R skills, depending on the application.

1.2.2 Installing and Updating R

Installing

As mentioned before, R is a free software package, which can be downloaded legally from the Internet page http://cran.r-project.org/bin.

Since R is a cross-platform software package, installing R on different operating systems will be explained. A full installation guide for all systems is available at http://cran.r-project.org/doc/manuals/R-admin.html.

Precompiled binary distributions

There are several ways of setting up R on a computer. On the one hand, for many operating systems, precompiled binary files are available. And on the other hand, for those who use other operating systems, it is possible to compile the programme from the source code.

- **Installing** R **under Unix**

  Precompiled binary files are available for the *Debian, RedHat, SuSe* and *Ubuntu* Unix distributions. They can be found on the CRAN website at http://cran.r-project.org/bin/linux.

- **Installing** R **under Windows**

  The binary version of R for Windows is located at http://cran.r-project.org/bin/windows.

  If an account with Administrator privileges is used, R can be installed in the *Program Files* path and all the optional registry entries are automatically set. Otherwise, there is only the possibility of installing R in the user files path. Recent versions of Windows ask for confirmation to proceed with installing a programme from an 'unidentified publisher'. The installation can be customised, but the default is suitable for most users.

  For further information, it is suggested to visit http://cran.r-project.org/bin/windows/base/rw-FAQ.html

- **Installing** R **under Mac**

  The current version of R for Mac OS is located at http://cran.r-project.org/bin/macosx/.

  The installation package corresponding to the specific version of the Mac OS must be chosen and downloaded. During the installation, the Installer will guide the user through the necessary steps. Note that this will require the password or login of an account with administrator privileges. The installation can be customised, but the default is suitable for most users. After the installation, R can be started from the application menu.

  For further information, it is suggested to visit http://cran.r-project.org/bin/macosx/RMacOSX-FAQ.html.

Updating

The best way to upgrade R is to uninstall the previous version of R, then install the new version and copy the old installed packages to the library folder of the new installation. Command `update.packages(checkBuilt = TRUE, ask = FALSE)` will update the packages for the new installation. Afterwards, any remaining data from the old installation can be deleted. Old versions of the software may be kept due to the parallel structure of the folders of the different installations. In cases where the user has a personal library, the contents must be copied into an update folder before running the update of the packages.

1.2.3 Packages

A package is a file, which may be composed of R scripts (for example functions) or dynamic link libraries (DLL) written in other languages, such as C or

`Fortran`, that gives access to more functions or data sets for the current session. Some packages are ready for use after the basic installation, others have to be downloaded and then installed when needed. On all operating systems, the function `install.packages()` can be used to download and install a package automatically through an available internet connection. Command `install.packages` may require to decide whether to compile and install sources if they are newer, then binaries. When installing packages manually, there are slight differences between operating systems.

Unix

Gzipped tar packages can be installed using the UNIX console by

```
R CMD INSTALL /your_path/your_package.tar.gz
```

Windows

In the `R GUI`, one uses the menu *Packages*.

- With an available internet connection, new packages can be downloaded and installed directly by clicking the *Install Packages* button. In this case, it is proposed to choose the CRAN mirror nearest to the user's location, and select the package to be installed.
- If the `.zip` file is already available on the computer, the package can be installed through *Install Packages from Zip files*.

Mac OS

There is a recommended *Package Manager* in the `R.APP GUI`. It is possible to install packages from the shell, but we suggest having a look at the FAQ on the CRAN website first.

All systems

Once a package is installed, it should be loaded in a session when needed. This ensures that the software has all the additional functions and datasets from this package in memory. This can be done through the commands

```
> library(package)              or    > require(package)
```

If the requested package is not installed, the function `library()` gives an error, while `require()` is designed for use inside of other functions and only returns `FALSE` and gives a warning.

The package will also be loaded as the second item in the system search path. Packages can also be loaded automatically if the corresponding code line is included in the `.Rprofile` file.

To see the installed libraries, the functions `library()` or `require()` are used.

```
> library()
```

To detach or unload a loaded package one uses.

```
> detach("package:name", unload = TRUE)
```

The function `detach()` can also be used to remove any R object from the search path. This alternative usage will be shown later in this chapter.

1.3   First Steps

After this first impression of what R is and how it works, the next steps are to see how it is used and to get used to it. In general, users should be aware of the **case sensitivity** of the R language.

It is also convenient to know that previously executed commands can be selected by the 'up arrow' on the keyboard. This is particularly useful for correcting typos and mistakes in commands that caused an error, or to re-run commands with different parameters.

1.3.1   "Hello World !!!"

As a first example, we will write some code that gives the output 'Hello World !!!' and a plot, see Fig. 1.1. There is no need to understand all the lines of the code now.

```
> install.packages("rworldmap")
> require(rworldmap)
> data("countryExData", envir = environment())
> mapCountryData(joinCountryData2Map(countryExData),
+    nameColumnToPlot  = "EPI_regions",
+    catMethod         = "categorical",
+    mapTitle          = "Hello World!!!",
+    colourPalette     = "rainbow",
+    missingCountryCol = "lightgrey",
+    addLegend         = FALSE)
```

1.3.2   Getting Help

Once R has been installed and/or updated, it is useful to have a way to get help. To open the primary interface to the help system, one uses

Hello World!!!

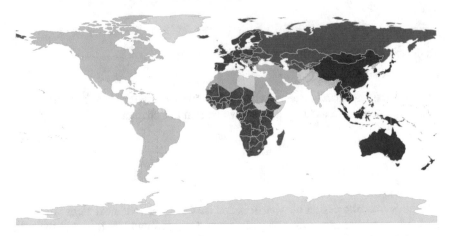

Fig. 1.1 "Hello World!!!" example in R. ⊙ BCS_HelloWorld

```
> help()
```

There are two ways of getting help for a particular function or command:

```
> help(function)                    and    > ? function
```

To find help about packages, one uses

```
> help(package = package)
```

which returns the help file that comes with the specific package. If the different proposals of help were not satisfying, one can try

```
> help.search("function name")           > ?? "function name"
```

to see all help subjects containing *"function name"*. Finally, under Windows and Mac OS, under the *Help* menu are several PDF manuals which provide thorough and detailed information. The same help can be reached with the function

```
> help.start()
```

An HTML version without pictures can be found on the CRAN website.
Examples for a particular topic can be found using

```
> example(topic)
```

1.3.3 Working Space

The current directory, where all pictures and tables are saved and from which all data is read by default, is known as the working directory. It can be found by `getwd()` and can be changed by `setwd()`.

```
> getwd()                              > setwd("your/own/path")
```

Each of the following functions returns a vector of character strings providing the names of the objects already defined in the current session.

```
> ls()                                 > objects()
```

To clear the R environment, objects can be deleted by

```
> rm(var1, var2)
```

The above line, for example, will remove the variables `var1` and `var2` from the working space. The next example will remove all objects defined in the current session and can be used to completely clean the whole working space.

```
> rm(list = ls())
```

The code below erases all variables, including the system ones beginning with a dot. Be cautious when using this command! Note that it has the same effect as the menu entry *Remove all objects* under Windows or *Clear Workspace* under Mac OS.

```
> rm(list = ls(all.names = TRUE))
```

However, we should always make sure that all previously defined variables are deleted or redefined when running new code, in order to be sure that there is no information left from the previous run of the programme which could affect the results. Therefore, it is recommended to have a line `rm(list = ls(all.names = TRUE))` in the beginning of each programme.

One saves the workspace as a *.Rdata* file in the specified working directory using the function `save.image()`, and saves the history of the commands in the *.R* format with `savehistory()`. Saving the workspace means keeping all defined variables in memory, so that the next time when R is in use, there is no need to define them again. If the history is saved, the variables will NOT be saved, whereas the commands defining them will be. So once the history is loaded, everything that was in the console should be compiled, but this can take a while for time-consuming calculations. The previously saved workspace and the history can be loaded with

```
> load(".Rdata")                       > loadhistory()
```

The `apropos('word')` returns the vector of functions, variables, etc., containing the argument *word*, as does `find('word')`, but with a different user interface. Without

going into details, the best way to set one's own search parameters is to consult the
help concerning these functions.

Furthermore, a recommended and very convenient way of writing programmes
is to split them into different modules, which might contain a list of definitions or
functions, in order not to mess up the main file. They are executed by the function

```
> source("my_module.r")
```

To write the output in a separate .txt file instead of the screen, one uses sink().
This file appears in the working directory and shows the full output of the session.

```
> sink("my_output.txt")
```

To place that output back on the screen, one uses

```
> sink()
```

The simplest ways to quit R are

```
> quit()                        or      > q()
```

1.4   Basics of the R Language

This section contains information on how R can be a useful software for all basic
mathematical and programming needs.

1.4.1   R as a Calculator

R may be seen as a powerful calculator which allows dealing with a lot of mathematical functions. Classical fundamental operations as presented in Tables 1.1, 1.2
and 1.3 are, of course, available in R.

In contrast to the classical calculator, R allows assigning one or more values to a
variable.

```
> a = pi + 0.5; a    # create variable a; print a
[1] 3.641593
```

Table 1.1 Fundamental operations

Function name	Example	Result
Addition	`1 + 2`	3
Subtraction	`1 - 2`	-1
Multiplication	`1 * 2`	2
Division	`1 / 2`	0.5
Raising to a power	`3^2`	9
Integer division	`5 %/% 2`	2
Modulo division	`5 %% 2`	1

Table 1.2 Basic functions

Function name	Example	Result
Square root	`sqrt(2)`	1.414214
Sine	`sin(pi)`	1.224606e-16
Cosine	`cos(pi)`	-1
Tangent	`tan(pi/4)`	1
Arcsine	`asin(pi/4)`	0.903339
Arccosine	`acos(0)`	1.570796
Arctangent	`atan(1)`	0.785398
Arctan(y/x)	`atan2(1, 2)`	0.463647
Hyperbolic sine	`sinh(1)`	1.175201
Hyperbolic cosine	`cosh(0)`	1
Hyperbolic tangent	`tanh(pi)`	0.9962721
Exponential	`exp(1)`	2.718282
Logarithm	`log(1)`	0

Table 1.3 Comparison relations

Meaning	Example	Result
Smaller	`5 < 5`	FALSE
Smaller or equal	`3 <= 4`	TRUE
Bigger	`7 > 2`	TRUE
Bigger or equal	`5 >= 5`	TRUE
Unequal	`2 != 1`	TRUE
Logical equal	`pi == acos(-1)`	TRUE

Transformations of numbers are implemented by the following functions.

```
> floor(a)                          > trunc(a)
[1] 3                               [1]   3
> floor(-a)                         > trunc(-a)
[1] -4                              [1] -3
> ceiling(a)                        > round(a)
[1]   4                             [1]   4
> ceiling(-a)                       > round(-a)
[1] -3                              [1] -4
> round(a, digits = 2)              > factorial(a)
[1] 3.64                            [1] 14.19451
```

`floor()` (`ceiling()`) returns the largest (smallest) integer that is smaller (larger) than the value of the given variable a, `trunc()` truncates the decimal part of a real-valued variable to obtain an integer variable. The function `round()` rounds a real-valued variable scientifically to an integer unless the argument `digits` is applied specifying the number of decimal places, in which case it scientifically rounds the given real number to that many decimal places. Scientific rounding of a real number rounds it to the closest integer, except for the case there the number after the predefined decimal place is exactly 5. For this case the closest even integer is returned. The function `factorial()`, which for an integer a returns $f(a) = a! = 1 \cdot 2 \cdot \ldots \cdot a$, works with real-valued arguments as well, by using the Gamma function

$$\Gamma(x) = \int_0^\infty t^{x-1} \exp(-t) \, dt,$$

implemented by `gamma(x+1)` in R.

1.4.2   Variables

Assigning variables

There are different ways to assign variables to symbols.

```
> a = pi + 0.5; a          # assign (pi + 0.5) to a
[1] 3.641593
> b = a; b                 # assign the value of a to b
[1] 3.641593
> d = e = 2^(1 / 2); d     # assign 2^(1 / 2) to e
>                          # and the value of e to d
[1] 1.414214
> e
[1] 1.414214
> f <- d; f                # assign the value of d to f
[1] 1.414214
> d -> g; g                # assign the value of d to g
[1] 1.414214
```

Be careful with using '=' for assigning, because the known argument, which defines the other, must be placed on the right side of the equals sign. The arrow assignment allows the following kind of constructions:

```
> h <- 4 -> j                # assign the value 4 to the variables h and j
```

These constructions should not be used extensively due to their evident lack of clarity. Note that variable names are case sensitive and must not begin with a digit or a period followed by a digit. Furthermore, names should not begin with a dot as this is common only for system variables. It is often convenient to choose names that contain the type of the specific variable, e.g. for the variable iNumber, the 'i' at the beginning indicates that the variable is of the type integer.

```
> iNumber = c(1, 2, 3, 4, 5, 6, 7, 8, 9, 10)
```

It is also useful to add the prefix 'L' to all local variables. For example, despite the fact that pi is a constant, one can reassign a different value to it. In order to avoid confusion, it would be convenient in this case to call the reassigned variable Lpi. We are not always following this suggestion in order to keep listings' lengths as short as possible.

It is also possible to define functions in a similar fashion.

```
> Stirling = function(n){sqrt(2 * pi * n) * (n / exp(1))^n}
```

This is the mathematical function known as *Stirling's formula* or *Stirling's approximation* for the factorial, which has the property

$$\frac{n!}{\sqrt{2\pi n} \cdot (n/e)^n} \rightarrow 1, \quad \text{as } n \rightarrow \infty.$$

For multi-output functions, see Sect. 1.4.6.

Working with variables

There are different *types* of variables in R. A brief summary is given in Table 1.4. Obviously each variable requires its own storage space, so during computationally intensive calculations one should pay attention to the choice of the variables. The types *numeric* and *double* are identical. Both are vectors of a specific length and store real valued elements. A variable of the type *logical* contains only the values TRUE and FALSE.

The command returning the type, i.e. the storage mode, of an object is typeof(), the possible values are listed in the structure TypeTable. Alternatively, the function class() can be used, which in turn returns the class of the object and is often used in the object-oriented style of programming.

```
> typeof(object name)              > class(object name)
```

A *character* variable consists of elements within quotation marks.

Table 1.4 Variable types

Variable type	Example of memory needed	Result
Numeric	`object.size(numeric(2))`	`56 bytes`
Logical	`object.size(logical(2))`	`48 bytes`
Character	`object.size(character(2))`	`104 bytes`
Integer	`object.size(integer(2))`	`48 bytes`
Double	`object.size(double(2))`	`56 bytes`
Matrix	`object.size(matrix(0, ncol = 2))`	`216 bytes`
List	`object.size(list(2))`	`96 bytes`

```
> a = character(length = 2); a
[1] "" ""
> a = c(exp(1), "exp(1)")
> class(a)
[1] "character"
> a
[1] "2.71828182845905" "exp(1)"
```

A description of the variable types *matrix* and *list* is given in Sects. 1.4.3 and 1.4.5 in more detail. To show, or *print*, the content of a, one uses the function `print()`.

```
> print(a)
```

'Unknown values' as a result of missing values is a common problem in statistics. To handle this situation, R uses the value NA. Every operation with NA will give an NA result as well. Note that NA is different from the value NaN, which is the abbreviation for 'not a number'. If R returns NaN, the underlying operation is not valid, which obviously has to be distinguished from the case in which the data is not available. A further output, that the reader should worry about, is Inf denoting infinity. It is difficult to work with such results, but R provides tools which modify the class of results, like NaN, and there are functions to help in transforming variables from one type into another. The functions as.character, as.double, as.integer, as.list, as.matrix, as.data.frame, etc., coerce their arguments to the function specific type. It is therefore possible to transform results like NaN or Inf to a preferable type, which is often done in programming with R.

In the example below, the function `paste()` automatically converts the arguments to strings, by using the function `as.character()`, and concatenates them into a single string. The option `sep` specifies the string character that is used to separate them.

```
> paste(NA, 1, "hop", sep = "@") # concatenate objects, separator @
[1] "NA@1@hop"
> typeof(paste(NA, 1, "hop", sep = "@"))
[1] "character"
```

Furthermore, one can check if an R object is finite, infinite, unknown, or of any other type. The function is.finite(*argument*) returns a Boolean object (a vector or matrix, if the input is a vector or a matrix) indicating whether the values are finite or not. This test is also available to test types, such as is.integer(x), to test whether x is an integer or not, etc.

```
> x = c(Inf, NaN, 4)
> is.finite(x)            # check if finite
[1] FALSE FALSE   TRUE
> is.nan(x)               # check if NaN (operation not valid)
[1] FALSE TRUE   FALSE
> is.double(x)            # check if type double
[1] TRUE
> is.character(x)         # check if type character
[1] FALSE
```

1.4.3   Arrays

A vector is a one-dimensional array of fixed length, which can contain objects of one type only. There are only three categories of vectors: *numerical*, *character* and *logical*.

Basic manipulations

Joining the values 1, π, $\sqrt{2}$ in a vector is done easily by the concatenate function, c.

```
> v = c(1, pi, sqrt(2)); v     # concatenate
[1] 1.000000 3.141593 1.414214
```

The ith element of the vector can be addressed using v[i].

```
> v = c(1.000000, 3.141593, 1.414214)
> v[2]                         # 2nd element of v
[1] 3.141593
```

The indexing of a vector starts from 1. If an element is addressed that does not exist, e.g. v[0], the error NA or numeric(0) is returned. A numerical vector may be *integer* if it contains only integers, *numeric* if it contains only real numbers, and *complex* if it contains complex numbers. The length of a vector object v is found through

```
> v = c(1.000000, 3.141593, 1.414214)
> length(v)                    # length of vector v
[1] 3
```

Be careful with this function, keeping in mind that it *always returns one value, even for multi-dimensional arrays*, so one should know the nature of the objects one is dealing with.

One easily applies the same transformation to all elements of a vector. One can calculate, for example, the *elementwise inverse* with the command ^ (-1). This is still the case for other objects, such as arrays.

```
> v = c(1.000000, 3.141593, 1.414214)
> d = v + 3; d
[1] 4.000000 6.141593  4.414214
> v^(-1)
[1] 1.0000000 0.3183099 0.7071068
> v * v^(-1)
[1] 1 1 1
```

There are a lot of other ways to construct vectors. The function array(x, y) creates an array of dimension y filled with the value x only. The function seq(x, y, by = z) gives a sequence of numbers from x to y in steps of z. Alternatively, the required length can be specified by option length.out.

```
> c(1, 2, 3)
[1] 1 2 3
> 1:3
[1] 1 2 3
> array(1:3, 6)
[1] 1 2 3 1 2 3
> seq(1, 3)
[1] 1 2 3
> seq(1, 3, by = 2)
[1] 1 3
> seq(1, 4, length.out = 5)
[1] 1.00 1.75 2.50 3.25 4.00
```

One can also use the rep() function to create a vector in which some values are repeated.

```
> v = c(1.000000, 3.141593, 1.414214)
> rep(v, 2)                    # the vector twice
[1] 1.00 3.14 1.41 1.00 3.14 1.41
> rep(v, c(2, 0, 1))           # 1st value twice, no 2nd value
>                              # 3rd value once
[1] 1.00 1.00 1.41
> rep(v, each = 2)             # each value twice
[1] 1.00 1.00 3.14 3.14 1.41 1.41
```

With the second command of the above code, R creates a vector in which the first value of v should appear two times, the second zero times, and the third only once. Note that if the second argument is not an integer, R takes the rounded value. In the last call, each element is repeated twice, proceeding element per element.

The names of the months, their abbreviations and all letters of the alphabet are stored in predefined vectors. The months can be addressed in the vector month.name[]. For their abbreviations, use month.abb[]. Letters are stored in letters[] and capital letters in LETTERS[].

```
> s = c(2, month.abb[2], FALSE, LETTERS[6]); s
[1] "2"      "Feb"   "FALSE"   "F"
> class(s)
[1] "character"
```

Note that if one element in a vector is of type `character`, then all elements in the vector are converted to `character`, since a vector can only contain objects of one type.

To keep only some specific values of a vector, one can use different methods of conditional selection. The first is to use logical operators for vectors in R: "!" is the logical NOT, "&" is the logical AND and "|" is the logical OR. Using these commands, it is possible to perform a conditional selection of vector elements. The elements for which the conditions are TRUE can then, for example, be saved in another vector.

```
> v = c(1.000000, 3.141593, 1.414214)
> v > 0                     # element greater 0
[1] TRUE TRUE TRUE
> (v != 1) & (v > 0)        # element not equal to 1 and greater 0
[1] FALSE TRUE TRUE
```

In the last example, the first value is bigger than zero, but equal to one, so FALSE is returned. This method may be a little bit confusing for beginners, but it is very useful for working with multi-dimensional arrays.

Multiple selection of elements of a vector may be done using another vector of indices as arguments in the square brackets.

```
> v = c(1.000000, 3.141593, 1.414214)
> v[c(1, 3)]                    # 1st and 3rd element
[1] 1.000000 1.414214
> w = v[(v != 1) & (v > 0)]; w  # save the specified elements in w
[1] 3.141593 1.414214
```

To eliminate specific elements in a vector, the same procedure is used as for selection, but a minus sign indicates the elements which should be removed.

```
> v = c(1.0000, 3.1416, 1.4142)     >
> v[-1]    # exclude first          > v[-c(1, 3)] # excl. 1st and 3rd
[1] 3.1416 1.4142                   [1] 3.141593
```

For a one-dimensional vector function, `which` returns the index or indices of specific elements.

```
> v = c(1.000000, 3.141593, 1.414214)
> which(v == pi)    # indices of elements that fulfill the condition
[1] 2
```

There are different functions for working with vectors. Extremal values are found through the functions `min` and `max`, which return the minimal and maximal values of the vector, respectively.

```
> v = c(1.0000, 3.1416, 1.4142)          >
> min(v)                                  > max(v)
[1] 1.000000                             [1] 3.141593
```

However, this can be done simultaneously by the function `range`, which returns a vector consisting of the two extreme values.

```
> v = c(1.000000, 3.141593, 1.414214)
> range(v)             # min and max value
[1] 1.000000 3.141593
```

Joining the function `which()` with `min` or `max`, one gets the function `which.min` or `which.max` that returns the index of the smallest or largest element of the vector, respectively, and is equivalent to `which(x == max(x))` and `which (x == min(x))`.

Quite often, the elements of a vector have to be sorted before one can proceed with further transformations. The simplest function for this purpose is `sort()`.

```
> x = c(4, 2, 5, 7, 1, 9, 0, 3)
> sort(x)              # values in increasing order
[1] 0 1 2 3 4 5 7 9
```

Being a function, it does not modify the original vector x. To get the coordinates of the elements that are in the sorted vector, we use the function `rank()`.

```
> x = c(4, 2, 5, 7, 1, 9, 0, 3)
> rank(x)              # rank of elements in increasing order
[1] 5 3 6 7 2 8 1 4
```

In this example, the first value of the result is '5'. This means that the first element in the original vector x[1] = 4 is in the fifth place in the ordered vector. The inverse function to `rank()` is `order()`, which states the position of the element of the sorted vector in the original vector, e.g. the smallest element in x is the seventh, the second smallest is the fifth, etc.

```
> x = c(4, 2, 5, 7, 1, 9, 0, 3)
> order(x)      # positions of sorted elements in the original vector
[1] 7 5 2 8 1 3 4 6
```

Replacing specific values in a vector is done with the function `replace()`. This function replaces the elements of x that are specified by the second argument by the values given in the third argument.

Table 1.5 Cumulative functions

Meaning	Implementation	Result
Sum	`cumsum(1:10)`	1 3 6 10 15 21 28 36 45 55
Product	`cumprod(1:5)`	1 2 6 24 120
Minimum	`cummin(c(3:1,2:0,4:2))`	3 2 1 1 1 0 0 0 0
Maximum	`cummax(c(3:1,2:0,4:2))`	3 3 3 3 3 3 4 4 4

```
> v = 1:10; v
[1]  1  2  3  4  5  6  7  8  9 10
> replace(v, v < 3, 12)      # replace all els. smaller than 3 by 12
[1] 12 12  3  4  5  6  7  8  9 10
> replace(v, 6, 12)          # replace the 6th element by 12
[1]  1  2  3  4  5 12  7  8  9 10
```

The second argument is a vector of indices for the elements to be replaced by the values. In the second line, all numbers smaller than 3 are to be replaced by 12, while in the last line, the element with index 6 is replaced by 12. Note again that functions do not change the original vectors, so that the last output does not show 1 and 2 replaced by 12 after the second command.

There are also a few more functions for vectors which are of further interest. The function `rev()` returns the elements in reversed order, and `sum()` gives the sum of all the elements in the vector.

```
> x = c(4, 2, 5, 7, 1, 9, 0, 3)
> rev(x)                # reverse the order of x
[1] 3 0 9 1 7 5 2 4
> sum(x)                # sum all elements of x
[1] 31
```

More sophisticated ways of cumulative summation of the elements of vectors are given in Table 1.5.

Vectors can also be considered as sets, and for this purpose there exist binary set operators, such as a `%in%` b, which gives the elements of a that are also in b. More advanced functions for working with sets are discussed in Chap. 3.

```
> a = 1:3                # 1 2 3
> b = 2:6                # 2 3 4 5 6
> a %in% b               # FALSE  TRUE   TRUE
> b %in% a               # TRUE TRUE FALSE FALSE FALSE
> a = c("A","B")         # "A" "B"
> b = LETTERS[2:6]       # "B" "C" "D" "E" "F"
> a %in% b               # FALSE  TRUE
> b %in% a               # TRUE FALSE FALSE FALSE FALSE
```

In algebra and statistics, matrices are fundamental objects, which allow summarising a large amount of data in a simple format. In R, matrices are only allowed to have one data type for their entries, which is their main difference from data frames, see Sect. 1.4.4.

Creating a matrix

There are many possible ways to create a matrix, as shown in the example below.
The function matrix() constructs matrices with specified dimensions.

```
> matrix(0, 2, 5) # zeros, 2x5
     [,1] [,2] [,3] [,4] [,5]
[1,]   0    0    0    0    0
[2,]   0    0    0    0    0

> matrix(1:12, nrow = 3)
     [,1] [,2] [,3] [,4]
[1,]   1    4    7   10
[2,]   2    5    8   11
[3,]   3    6    9   12

> matrix(1:6, nrow = 2, byrow = TRUE)
     [,1] [,2] [,3]
[1,]   1    2    3
[2,]   4    5    6
```

In the third matrix, in the above example, the argument byrow = TRUE indicates that the filling must be done by rows, which is not the case in the second
matrix, where the matrix was filled by columns (*column-major storage*), the function as.vector(*matrix*) converts a matrix into a vector. If the matrix has more
than one row or column, the function concatenates the columns into a vector. One
can also construct diagonal matrices using diag(), see Sect. 2.1.1.

Another way to transform a given vector into a matrix with specified dimensions
is the function dim(). The function t() is used to transpose matrices.

```
> m       = 1:6                    > t(m)           # transpose m
> dim(m) = c(2, 3); m                  [,1] [,2]
     [,1] [,2] [,3]             [1,]   1    2
[1,]   1    3    5              [2,]   3    4
[2,]   2    4    6              [3,]   5    6
```

Coupling vectors using the functions cbind() (column bind) and rbind() (row
bind) joins vectors column-wise or row-wise into a matrix.

```
> x = 1:6
> y = LETTERS[1:6]
> rbind(x, y)      # bind vectors row-wise
  [,1] [,2] [,3] [,4] [,5] [,6]
x "1"  "2"  "3"  "4"  "5"  "6"
y "A"  "B"  "C"  "D"  "E"  "F"
```

The functions col and row return the column and row indices of all elements of
the argument, respectively.

```
> m = matrix(1:6, ncol = 3)          > m = matrix(1:6, ncol = 3)
> col(m)   # column-indices          > row(m)   # row-indices
     [,1] [,2] [,3]                        [,1] [,2] [,3]
[1,]    1    2    3                   [1,]    1    1    1
[2,]    1    2    3                   [2,]    2    2    2
```

The procedure to extract an element or submatrix uses a syntax similar to the syntax for vectors. In order to extract a particular element, one uses m[row index, column index]. As a reminder, in the example below, 10 is the second element of the fifth column, in accordance with the standard mathematical convention.

```
> k = matrix(1:10, 2, 5); k   # create a matrix
[,1] [,2] [,3] [,4] [,5]
[1,]    1    3    5    7    9
[2,]    2    4    6    8   10

> k [2, 5]                    # select element in row 2, column 5
[1] 10
```

One can combine this with row() and col() to construct a useful tool for the conditional selection of matrix elements. For example, extracting the diagonal of a matrix can be done with the following code.

```
> m = matrix(1:6, ncol = 3)
> m[row(m) == col(m)]         # select elements [1, 1]; [2, 2]; etc.
[1] 1 4
```

The same result is obtained by using the function diag(m). To better understand the process, note that the command row(m) == col(m) creates just the Boolean matrix below and all elements with value TRUE are subsequently selected.

```
> row(m) == col(m)            # condition (row index = column index)
     [,1]  [,2]   [,3]
[1,]  TRUE FALSE FALSE
[2,] FALSE  TRUE FALSE
```

This syntax can also be used to select whole rows or columns.

```
> y = matrix (1:16, ncol = 4, nrow = 4); y
     [,1] [,2] [,3] [,4]
[1,]    1    5    9   13
[2,]    2    6   10   14
[3,]    3    7   11   15
[4,]    4    8   12   16
> y[2, ]         # second row
[1]  2  6 10 14
> y[, 2]         # second column
[1] 5 6 7 8
> y[2]           # second element (column-wise)
[1] 2
> y[1:2, 3:4]    # several rows and columns
     [,1] [,2]
[1,]    9   13
[2,]   10   14
```

The first command selects the second row of y. The second command selects the 2nd column. The third line considers the matrix as a succession of column vectors and gives, according to this construction, the second element. The last call selects a range of rows and columns.

Many functions can take matrices as an argument, such as sum() or product(), which will calculate the sum or product of all elements in the matrix, respectively. The functions colSums() and rowSums() can be used to calculate the column-wise or row-wise sums. All classical binary operators are implemented element-by-element. This means that, for example, x * y returns the Kronecker product, not the classical matrix product discussed in Sect. 2.1 on matrix algebra.

One can assign names to the rows and columns of a matrix using the function dimnames(). Alternatively, the column and row names can be assigned separately by colnames() and rownames(), respectively.

```
> A            = matrix(1:20, ncol = 5, nrow = 4)
> dimnames(A)  = list(letters[4:7], letters[5:9]) # name dimensions
> A
  e f  g  h  i
d 1 5  9 13 17
e 2 6 10 14 18
f 3 7 11 15 19
g 4 8 12 16 20
> A[2, 2]
[1] 6
> A["b", "b"]
[1] 6
```

This leads directly to another very useful format in R: the data frame.

1.4.4 Data Frames

A data frame is a very useful object, because of the possibility of collecting data of different types (numeric, logical, factor, character, etc.). Note, however, that all elements must have the same length. The function data.frame() creates a new data frame object. It contains several arguments, as the column names can be directly specified with data.frame(..., row.names = c(), ...). A further possibility for creating a data frame is to convert it from a matrix with the as.data.frame(matrix name) function.

Basic manipulations

Consider the following example, which constructs a data frame.

```
> cities     = c("Berlin", "New York", "Paris", "Tokyo")
> area       = c(892, 1214, 105, 2188)
> population = c(3.4, 8.1, 2.1, 12.9)
> continent  = factor(c("Europe", "North America", "Europe", "Asia"))
> myframe    = data.frame(cities, area, population, continent)
> is.data.frame(myframe) # check if object is a dataframe
[1] TRUE
> rownames(myframe) = c("Berlin", "New York", "Paris", "Tokyo")
```

```
> colnames(myframe) = c("City", "Area", "Pop.", "Continent")
> myframe
     City Area Pop.       Continent
Berlin      Berlin  892  3.4          Europe
New York New York 1214  8.1 North America
Paris        Paris  105  2.1          Europe
Tokyo        Tokyo 2188 12.9            Asia
```

Note that if we defined the above data frame as a matrix, then all elements would be converted to type `character`, since matrices can only store one data type.

`data.frame()` automatically calls the function `factor()` to convert all character vectors to `factors`, as it does for the `Continent` column above, because these variables are assumed to be indicators for a subdivision of the data set. To perform data analysis (e.g. principal component analysis or cluster analysis, see Chap. 8), numerical expressions of character variables are needed. It is therefore often useful to assign ordered numeric values to character variables, in order to perform statistical modelling, set the correct number of degrees of freedom, and customise graphics. These variables are treated in R as factors. As an example, a new variable is constructed, which will be added to the data frame "*myframe*". Three position categories are set, according to the proximity of each city to the sea: `Coastal` ('0'), `Middle` ('1') and `Inland` ('2'). These categories follow a certain order, with `Middle` being in between the others, which needs to be conveyed to R.

```
> e         = c(2, 0, 2, 0)                          # code info. in e
> f         = factor(e, level = 0:2)                 # create factor f
> levels(f) = c("Coastal", "Middle", "Inland"); f # with 3 levels
[1] Inland  Coastal Inland  Coastal
Levels: Coastal Middle Inland
> class(f)
[1] "factor"
> as.numeric(f)
[1] 3 1 3 1
```

The variable f is now a `factor`, and levels are defined by the function `levels()` in the 3rd line in decreasing order of the proximity to the sea. When sorting the variable, R will now follow the order of the levels. If the position values were simply coded as *string*, i.e. `Coastal`, `Middle` and `Inland`, any sorting would be done *alphabetically*. The first level would be `Coastal`, but the second `Inland`, which does not follow the inherited order of the category.

The function `as.numeric()` extracts the numerical coding of the levels and the indexation begins now with 1.

```
> myframe                = data.frame(myframe, f)
> colnames(myframe)[5] = "Prox.Sea" # name 5th column
> myframe
     City Area Pop.       Continent Prox.Sea
Berlin      Berlin  892  3.4          Europe   Inland
New York New York 1214  8.1 North America  Coastal
Paris        Paris  105  2.1          Europe   Inland
Tokyo        Tokyo 2188 12.9            Asia  Coastal
```

The column names for columns 1 to 4 are the ones that were assigned before, since `myframe` is used in the call of `data.frame()`. Note that one should not use names with spaces, e.g. `Sea.Env.` instead of `Sea. Env`. To add columns or rows to a data frame, one can use the same functions as for matrices, or the procedure described below.

```
> myframe = cbind(myframe, "Language.Spoken"=
+    c("German", "English", "French", "Japanese"))
> myframe
         City Area Pop.      Continent Prox.Sea Language.Spoken
Berlin     Berlin  892  3.4     Europe    Inland         German
New York New York 1214  8.1 North America Coastal        English
Paris       Paris  105  2.1     Europe    Inland          French
Tokyo       Tokyo 2188 12.9       Asia   Coastal        Japanese
```

There are several ways of addressing one particular column by its name: *myframe$Pop.*, *myframe[, 3]*, *myframe[, "Pop."]*, *myframe ["Pop."]*. All these commands except the last return a *numeric vector*. The last command returns a *data frame*.

```
> myframe$Pop.      # select only population column
[1]  3.4  8.1  2.1 12.9
> myframe["Pop."] # population column as dataframe
          Pop.
Berlin    3.4
New York  8.1
Paris     2.1
Tokyo    12.9
> myframe[3] == myframe["Pop."]
          Pop.
Berlin    TRUE
New York  TRUE
Paris     TRUE
Tokyo     TRUE
```

The output of the above code is a data frame and, therefore, can not be indexed like a vector. One uses $ notation similar to addressing fields of objects in the C++ programming language.

```
> myframe[2, 3]     # select 3rd entry of 2nd row
[1] 8.1
> myframe[2, ]      # select 2nd row
         City Area Pop.      Continent Prox.Sea Language.Spoken
New York New York 1214  8.1 North America Coastal        English
```

Long names for data frames and the contained variables should be avoided, because the source code becomes very messy if several of them are called. This can be solved by the function `attach()`. Attached data frames will be set to the search path and the included variables can be called directly. Any R object can be attached. To remove it from the search path, one uses the function `detach()`.

```
> rm(area)          # remove var. "area" to avoid confusion
> attach(myframe)   # attach dataframe "myframe"
> Area              # specify column Area in attached frame
[1]   892 1214  105 2188
> detach(myframe)
```

If two-word names are used, it is advised to label the data frame or variable with a block name, so that the two words in the name are connected with a dot or an underline, e.g. `Language.Spoken`. This avoids having to put names in quotes.

One of the easiest ways to edit a data frame or a matrix is through interactive tables, called by the `edit` function. Note that the `edit()` function does not allow changing the original data frame.

```
> edit(myframe)
```

If the modifications are to be saved, the function `fix()` is employed. It opens a table like `edit()`, but the changes in the data are stored.

```
> fix(myframe)
```

A data frame as a database

Furthermore, R provides the possibility of selecting subsets of a data frame by using the logical operators discussed in Sects. 1.4.1 and 1.4.3: $<$, $>$, $=<$, $>=$, $==$, $!=$, &, | and !.

```
> myframe[(myframe$Language.Spoken == "French") |
+    (myframe$Pop. > 10), -1]
       Area Pop. Continent Prox.Sea Language.Spoken
Paris   105  2.1    Europe   Inland          French
Tokyo  2188 12.9      Asia  Coastal        Japanese
> myframe[, -c(1, 2, 3, 5)] # select all except specified columns
                Continent Language.Spoken
Berlin             Europe          German
New York    North America         English
Paris              Europe          French
Tokyo                Asia        Japanese
```

The first command of the last listing selects both the cities in which French is spoken or the cities with more than 10 million inhabitants. The second command selects only the first, fourth and sixth variables for display. As explained above, the individual data, as well as rows and columns, can be addressed using the square brackets. If no variable is selected, i.e. `[ , ]`, all information about the observations is kept.

The following functions are also helpful for conditional selections from data frames. The function `subset()`, which performs conditional selection from a data frame, is frequently used when only a subset of the data is used for the analysis.

```
> subset(myframe, Area > 1000)
              City Area Pop.     Continent Prox.Sea Language.Spoken
New York  New York 1214  8.1 North America  Coastal         English
Tokyo        Tokyo 2188 12.9          Asia  Coastal        Japanese
```

A conditional transformation of the data frames, by adding a new variable which is a function of others, is done by using the function `transform()`. As an example, a new variable `Density` is added to our data frame.

```
> transform(myframe[, -c(1, 4, 5)], Density = Pop. * 10^6 / Area)
         Area Pop. Language.Spoken   Density
Berlin    892  3.4          German  3811.659
New York 1214  8.1         English  6672.158
Paris     105  2.1          French 20000.000
Tokyo    2188 12.9        Japanese  5895.795
```

Another way to extract data according to the values is based on addressing specific variables. In the next example, the interest is in the cumulative area of cities that are not inland.

```
> Area.Seasiders = myframe$Area[myframe$Prox.Sea == "Middle"
+      | myframe$Prox.Sea == "Coastal"]
> Area.Seasiders
[1] 1214 2188
> sum(Area.Seasiders)
[1] 3402
```

The important technique of sorting the data frame is illustrated below. Remember that order() sorts the elements and returns their ranks in the original vector. The optional argument partial specifies the columns for subsequent ordering, if necessary. It is used to order groups of data according to one column and order the values in each group according to another column.

```
> myframe[order(myframe$Pop., partial = myframe$Area), ]
             City Area Pop.    Continent Prox.Sea Language.Spoken
Paris       Paris  105  2.1       Europe   Inland          French
Berlin     Berlin  892  3.4       Europe   Inland          German
New York New York 1214  8.1 North America Coastal         English
Tokyo       Tokyo 2188 12.9         Asia  Coastal        Japanese
```

1.4.5   Lists

Lists are very flexible objects which, unlike matrices and data frames, may contain variables of different types and lengths.

The simplest way to construct a list is by using the function list(). In the following example, a string, a vector and a function are joined into one variable.

```
> a = c(2, 7)
> b = "Hello"
> d = list(example = Stirling, a, end = b)
> d
$example
function(x){
  sqrt(2 * pi * x) * (x / exp(1))^x
}

[[2]]
[1] 2 7

$end
[1] "Hello"
```

Another way to join these into a list is through a *vector* construction.

```
> z = c(Stirling, a)
> typeof(z)
[1] "list"
```

To address the elements of a list object, one again uses '$', the same syntax as for a data frame.

```
> d$end
[1] "Hello"
```

A list can be transformed into a 1-element list, i.e. a list of length 1, using `unlist`. In this example, the element `[[2]]` of list `d` is split into two elements, each of length 1.

```
> unlist(d)   # transform to list with elements of length 1
$example
function(x){
  sqrt(2 * pi * x) * (x / exp(1))^x
}


[[2]]
[1] 2

[[3]]
[1] 7

$end
[1] "Hello"
```

One of the possible ways of converting objects is to use the function `split()`. This returns a list of the split objects with separations according to the defined criteria.

```
> split(myframe, myframe$Continent)
$Asia
          City Area Pop. Continent Prox.Sea Language.Spoken
Tokyo Tokyo 2188 12.9      Asia    Coastal         Japanese


$Europe
          City Area Pop. Continent Prox.Sea Language.Spoken
Berlin Berlin  892  3.4    Europe   Inland           German
Paris   Paris  105  2.1    Europe   Inland           French

$`North America`
              City Area Pop.     Continent Prox.Sea Language.Spoken
New York New York 1214  8.1 North America  Coastal          English
```

In the above example, the data frame `myframe` is split into elements according to its column `Continent` and transformed into a list.

1.4.6 Programming in **R**

Functions

R has many programming capabilities, and allows creating powerful routines with functions, loops, conditions, packages and objects. As in the Stirling example, `args()` is used to receive a list of possible arguments for a specific function.

```
> args(data.frame)    # list possible arguments and default values
function(..., row.names = NULL, check.rows = FALSE, check.names
    = TRUE, stringsAsFactors = default.stringsAsFactors())
NULL
```

This command provides a list of all arguments that can be used in the function, including the default settings for the optional ones, which have the form *optional argument = setting value.*

Below a simple function is presented, which returns the list $\{a \cdot \sin(x), a \cdot \cos(x)\}$. The arguments a and x are defined in round brackets. We can define functions with optional arguments that have default values, in this example, a = 1.

```
> myfun = function(x, a = 1){    # define function
+     r1 = a * sin(x)
+     r2 = a * cos(x)
+     list(r1, r2)
+ }
> myfun(pi / 2)                  # apply to pi / 2, a = default
[[1]]
[1] 1

[[2]]
[1] 6.123234e-17
```

Note that if no `return(result)` operator is given at the end of the function body, then the last created object will be returned.

Loops and conditions

The family of these operators is a powerful and useful tool. However, in order to perform well, they should be used wisely. Let us start with the '*if*' condition.

```
> x = 1
> if(x == 2){print("x == 2")}
> if(x == 2){print("x == 2")}else{print("x != 2")}
[1] "x != 2"
```

The first programme is only an `if` condition, whereas the second is extended by the `else` command, which provides an alternative command in case the condition is not realised. More advanced, but more unusual in syntax, is the function `ifelse(`*boolean check, if-case, else-case*`)`. It is used mainly in advanced frameworks, to simplify code in which several `ifelse` constructions are embedded within each other.

Furthermore, `for` and `while` are very useful functions for creating loops, but are best avoided in case of large sample sizes and extensive computations, since they work very slowly. The difference between the functions is that `for` applies the computation for a defined range of integers and `while` carries out the computation until a certain condition is fulfilled. One may also use `repeat`, which will repeat the specified code until it reaches the command `break`. One must be careful to include a break rule or the loop will repeat infinitely.

```
> x = numeric(1)
> # for i from 1 to 10, the i-th element of x takes value i
> for(i in 1:10) x[i] = i
> x
 [1]  1  2  3  4  5  6  7  8  9 10
> # as long as i < 21, set i-th element equal to i and increase i by 1
> i = 1
> while(i < 21){
+     x[i] = i
+     i    = i + 1
+ }
> x
 [1]  1  2  3  4  5  6  7  8  9 10 11 12 13 14 15 16 17 18 19 20
> # remove the first element of x, stop when x has length 1
> repeat{
+     x = x[-1]
+     if(length(x) == 1) break
+ }
> x
[1] 20
```

As an alternative to loops, R provides the functions `apply()`, `sapply()` (for multivariate case `mapply()`) and `lapply()`. In many cases, using these functions will improve the computational time significantly compared to the above loop functions. The function `apply()` applies a deterministic function to the rows or to the columns of a matrix. The first argument specifies the matrix, the second argument determines whether the third argument is applied to rows or columns, and the third argument specifies the function.

```
> A = matrix(1:24, 12, 2, byrow = TRUE)
> apply(A, 2, mean) # apply mean to columns separately
[1] 12 13
> apply(A, 2, sum)  # column-wise sum
[1] 144 156
> apply(A, 1, mean) # row-wise mean
 [1]  1.5  3.5  5.5  7.5  9.5 11.5 13.5 15.5 17.5 19.5 21.5 23.5
> apply(A, 1, sum)  # row-wise sum
 [1]  3  7 11 15 19 23 27 31 35 39 43 47
```

The functions `lapply()` and `sapply()` are of a more general form, since the function is applied to each element of the list object and returns a list; and `sapply` returns a numeric vector or a matrix, if appropriate. So if the class of the object is *matrix* or *numeric*, the `sapply` function is preferred, but this function takes longer, as it applies `lapply` and converts the result afterwards. If a more general object, e.g. a list of objects, is used, the `lapply` function is more appropriate.

```
> A = matrix(1:24, 12, 2, byrow = TRUE)
> # apply function sin() to every element, return numeric vector
> sapply(A[1, ], sin)
[1] 0.8414710 0.9092974
> class(sapply(A[1:4, ], sin))
[1] "numeric"

> # apply function sin() to every element, return list
> lapply(A[1, ], sin)
[[1]]
[1] 0.841471

[[2]]
[1] 0.9092974
> class(lapply(A[1:4, ], sin))
[1] "list"
```

There is one more useful function, called `tapply()`, which applies a defined function to each cell of a ragged array. The latter is made from non-empty groups of values given by a unique combination of the levels of certain factors. Simply speaking, `tapply()` is used to break an array or vector into different subgroups before applying the function to each subgroup. In the example below, a matrix A is examined, which could contain the observations 1–12 of individuals from group 1, 2 or 3. Our intention is to calculate the mean for each group separately.

```
> g = c(rep(1, 4), rep(2, 4), rep(3, 4)) # create vector "group ID"
> A = cbind(1:12, g)                      # observations and group ID
> tapply(A[, 1], A[, 2], mean)            # apply function per group
1    2    3
2.5  6.5 10.5
```

Finally, the `switch()` function may be seen as the highlight of R's built-in programming functions. The function `switch(i, expression1, expression2,...)` chooses the i-th expression in the given expression arguments. It works with numbers, but also with character chains to specify the expressions. This can be used to simplify code, e.g. by defining a function that can be called to perform different computations.

```
> rootsquare = function(x, type){ # define function for ^2 or ^(0.5)
+    switch (type, square = x * x, root = sqrt(x))
+ }
> rootsquare(10, "square")       # apply "square" to argument 10
[1] 100
> rootsquare(10, 1)              # first is equivalent to "square"
[1] 100
> rootsquare(10, "root")         # apply "root" to argument 10
[1] 3.162278
> rootsquare(10, 2)
[1] 3.162278
> rootsquare(10, "ROOT")         # apply "ROOT" (not defined)
[1] NULL
```

It is sometimes useful to compare the efficiency of two different commands in terms of the time they need to be computed, which can be done by the function system.time(). This function returns three values: *user* is the CPU time charged for the execution of the user instructions of R, i.e. the processing of the functions. The *system time* is the CPU time used by the system on behalf of the calling process. In sum, they give the total amount of CPU time. *Elapsed time* is the total real time passed for the user. Since CPU processes can be run simultaneously, there is no clear relation between total CPU time and elapsed time.

```
> x = c(1:500000)
> system.time(for(i in 1:200000) {x[i] = rnorm(1)})
user  system elapsed
0.892   0.033   0.925
> system.time(x <- rnorm(200000))
user  system elapsed
0.017   0.000   0.017
```

Here the function rnorm(*x*) is used, which simulates from the normal distribution, see Sect. 4.3. Note that the hardwired rnorm() is faster than the for loop.

1.4.7 Date Types

R provides full access to current date and time values through the functions

```
> Sys.time()                                    > date()
```

The function as.Date() is used to format data from another source to dates that R can work with. When reading in dates, it is important to specify the date format, i.e. the order and delimiter. A list of date formats that can be converted by R via the appropriate conversion specifications can be found in the help for strptime. One can also change the format of the dates in R.

```
> dates = c("23.05.1984", "2001/01/01", "May 3, 1256")
> # read dates specifying correct format
> dates1 = as.Date(dates[1], "%d.%m.%Y"); dates1
[1] "1984-05-23"
> dates2 = as.Date(dates[2], "%Y/%m/%d"); dates2
[1] "2001-01-01"
> dates3 = as.Date(dates[3], "%B %d,%Y"); dates3
[1] "1256-05-03"
> dates.a = c(dates1, dates2, dates3)
> format(dates.a, "%m.%Y")   # delimiter "." and month/year only
[1] "05.1984" "01.2001" "05.1256"
```

Note that the function as.Date is not only applicable to *character strings, factors* and *logical NA*, but also to objects of types *POSIXlt* and *POSIXct*. The last two objects represent calendar dates and times, where *POSIXct* denotes the UTC timezone as a numeric vector and *POSIXlt* gives a list of vectors including seconds, minutes, hours, etc.

The difference between two dates is calculated by `difftime()`.

```
> dates.a = as.Date(c("1984/05/23", "2001/01/01", "1256/05/03"))
> difftime(Sys.time(), dates.a)
Time differences in days
[1]   11033.473    5332.473 276949.473
```

The functions `months()`, `weekdays()` and `quarters()` give the month, week-day and quarter of the specified date, respectively.

```
> dates.a = as.Date(c("1984/05/23", "2001/01/01", "1256/05/03"))
> months(dates.a)
[1] "May"        "January" "May"
> weekdays(dates.a)
[1] "Wednesday" "Saturday"   "Wednesday"
> quarters(dates.a)
[1] "Q2" "Q1" "Q2"
```

1.4.8   Reading and Writing Data from and to Files

For statisticians, software must be able to easily handle data without restrictions on its format, whether it is 'human readable' (such as `.csv`, `.txt`), in binary format (SPSS, STATA, Minitab, S-PLUS, SAS (export libs)) or from relational databases.

Writing data

There are some useful functions for writing data, e.g. the standard `write.table()`. Its often used options include `col.names` and `row.names`, which specify whether row or column names are written to the data file, as well as `sep`, which specifies the separator to be used between values.

```
> write.table(myframe,  "mydata.txt")
> write.table(Orange, "example.txt",
+    col.names = FALSE, row.names = FALSE)
> write.table(Orange, "example2.txt", sep="\t")
```

The first command creates the file *mydata.txt* in the working directory of the data frame *myframe* from Sect. 1.4.4, the second specifies that the names for columns and rows are not defined, and the last one asks for *tab separation* between cells.

The functions `write.csv()` and `write.csv2()` are both used to create Excel-compatible files. They differ from `write.table()` only in the default definition of the decimal separator, where `write.csv()` uses '.' as a decimal separator and ',' as the separator between columns in the data. Function `write.csv2()` uses ',' as decimal separator and ';' as column separator.

```
> data = write.csv("file name")  # decimal ".", column separator ","
> data = write.csv2("file name") # decimal ",", column separator ";"
```

Reading data

R supplies many built-in data sets. They can be found through the function `data()`, or, less efficiently, through `objects(package:datasets)`. In any case, we can load the pre-built data sets library using library("datasets"), where the quotation marks are optional. Many packages bring their own data, so for many examples in this book, packages will be loaded in order to work with their included data sets. To check whether a data set is in a package, data(package = "package name") is used.

Moreover, we can import .txt files, with or without header.

```
> data = read.table("mydata.txt", header = TRUE)
```

The default option is header = FALSE, indicating that there is no description in the first row of each column. Information about the data frame is requested by the functions names(), which states the column names, and str(), which displays the structure of the data.

```
> names(data)
[1] "City"            "Area"              "Pop."          "Continent"
[5] "Prox.Sea"        "Language.Spoken"
> str(data)
'data.frame':    4 obs. of  6 variables:
 $ City            : Factor w/ 4 levels "Berlin","New York",..:1 2 3 4
 $ Area            : int   892 1214 105 2188
 $ Pop.            : num   3.4 8.1 2.1 12.9
 $ Continent       : Factor w/ 3 levels "Asia","Europe",..:2 3 2 1
 $ Prox.Sea        : Factor w/ 2 levels "Coastal","Inland":2 1 2 1
 $ Language.Spoken: Factor w/ 4 levels "English","French",..:3 1 2 4
```

The function head() returns the first few rows of an object, which can be used to check whether there is a header. To do this, it is important to know that the first row is generally a header when it has one column less than the second row.

In some cases, the separation between the columns will not follow any of the standard formats. We can then use option sep to manually specify the column separator.

```
> data = read.table("file name", sep = "\t")
```

Here we specified manually that there is a tab character between the variables. Without a correctly specified separator, R may read all the lines as a single expression.

Missing values are represented by NA in R, but different programmes and authors use other symbols, which can be defined in the function. Suppose, for example, that the NA values were denoted by 'missing' by the creator of the dataset.

```
> data = read.table("file name", na.strings = "missing")
```

To import data in .csv (comma-separated list) format, e.g. from Microsoft Excel, the functions read.csv() or read.csv2() are used. They differ from each other in the same way as the functions write.csv() or write.csv2() discussed above.

To import or even write data in the formats of statistic software packages such as STATA or SPSS, the package foreign provides a number of additional functions. These functions are named read. plus the data file extension, e.g. read.dta() for STATA data.

To read data in the most general way from any file, the function scan("*file name*") is used. This function is more universal than read.table(), but not as simple to handle. It can be used to read columnar data or read data into a list.

It is possible to have the user choose interactively between several options by using the function menu (). This function shows a list of options from which the user can choose by entering the value or its index number, and gives as output the list rank. With the option graphics = TRUE, the list is shown in a separate window.

```
> menu(c("abc", "def"), title = "Enter value")
Enter value

1: abc
2: def

Selection: def
[1] 2
> menu(c("abc", "def"), graphics = TRUE, title = "Enter value")
```

Chapter 2
Numerical Techniques

The general who wins a battle makes many calculations in the temple before an attack.

— Sun Tzu, *The Art of War*

With more and more practical problems of applied mathematics appearing in different disciplines, such as chemistry, biology, geology, management and economics, to mention just a few, the demand for numerical computation has considerably increased. These problems frequently have no analytical solution or the exact result is time-consuming to derive. To solve these problems, numerical techniques are used to approximate the result. This chapter introduces matrix algebra, numerical integration, differentiation and root finding.

2.1 Matrix Algebra

Matrix algebra is a fundamental concept for applying and understanding numerical methods, therefore the beginning of this section introduces the basic characteristics of matrices, their operations and their implementation in R. Thereafter, other operations, such as the inverse, including the inverse both of non-singular and of singular matrices, the norms containing the vector norms and the matrix norms, the calculation of eigenvalues and eigenvectors and different types of matrix decompositions are presented. Some theorems and accompanying examples computed in R are also provided.

© Springer International Publishing AG 2017
W.K. Härdle et al., *Basic Elements of Computational Statistics*,
Statistics and Computing, DOI 10.1007/978-3-319-55336-8_2

2.1.1 Characteristics of Matrices

A matrix $A_{(n \times p)}$ is a system of numbers with n rows and p columns:

$$A = \begin{pmatrix} a_{11} & a_{12} & \cdots & a_{1p} \\ a_{21} & a_{22} & \cdots & a_{2p} \\ \vdots & \vdots & \ddots & \vdots \\ a_{n1} & \cdots & \cdots & a_{np} \end{pmatrix} = (a_{ij}).$$

Matrices with one column and n rows are column vectors, and matrices with one row and p columns are row vectors. The following R code produces (3×3) matrices A and B with the numbers from 1 to 9 and from 0 to -8, respectively. The matrices are filled by rows if byrow = TRUE.

```
> # set matrices A and B
> A = matrix(1:9, nrow = 3, ncol = 3, byrow = TRUE); A
     [,1] [,2] [,3]
[1,]    1    2    3
[2,]    4    5    6
[3,]    7    8    9
> B = matrix(0:-8, nrow = 3, ncol = 3, byrow = FALSE); B
     [,1] [,2] [,3]
[1,]    0   -3   -6
[2,]   -1   -4   -7
[3,]   -2   -5   -8
```

There are several special matrices that are frequently encountered in practical and theoretical work. Diagonal matrices are special matrices where all off-diagonal elements are equal to 0, that is, $A_{(n \times p)}$ is a diagonal matrix if $a_{ij} = 0$ for all $i \neq j$. The function diag() creates diagonal matrices (square or rectangular) or extracts the main diagonal of a matrix in R.

```
> A = matrix(1:9, nrow = 3, ncol = 3, byrow = TRUE)
> diag(x = A)                             # extract diagonal
[1] 1 5 9
> diag(3)                                 # identity matrix
     [,1] [,2] [,3]
[1,]    1    0    0
[2,]    0    1    0
[3,]    0    0    1
> diag(2, 3)                              # 2 on diag, 3x3
     [,1] [,2] [,3]
[1,]    2    0    0
[2,]    0    2    0
[3,]    0    0    2
> diag(c(1, 5, 9, 13), nrow = 3, ncol = 4) # 3x4
     [,1] [,2] [,3] [,4]
[1,]    1    0    0    0
[2,]    0    5    0    0
[3,]    0    0    9    0
> diag(2, 3, 4)                           # 3x4, 2 on diagonal
     [,1] [,2] [,3] [,4]
[1,]    2    0    0    0
[2,]    0    2    0    0
```

```
[3,]    0    0    2    0
```

As seen from the listing above, the argument x of `diag()` can be a matrix, a vector, or a scalar. In the first case, the function `diag()` extracts the diagonal elements of the existing matrix, and in the remaining two cases, it creates a diagonal matrix with a given diagonal or of given size.

Rank

The rank of $\mathcal{A}$ is denoted by $\text{rank}(\mathcal{A})$ and is the maximum number of linearly independent rows or columns. Linear independence of a set of h rows a_j means that $\sum_{j=1}^{h} c_j a_j = 0_p$ if and only if $c_j = 0$ for all j. If the rank is equal to the number of rows or columns, the matrix is called a full-rank matrix. In R the rank can be calculated using the function `qr()` (which does the so-called QR decomposition) with the object field `rank`.

```
> A = matrix(1:9, nrow = 3, ncol = 3, byrow = TRUE)
> qr(A)$rank # rank of matrix A
[1] 2
```

The matrix A is not of full rank, because the second column can be represented as a linear combination of the first and third columns:

$$\begin{pmatrix} 2 \\ 5 \\ 8 \end{pmatrix} = \frac{1}{2} \begin{pmatrix} 1+3 \\ 4+6 \\ 7+9 \end{pmatrix} \tag{2.1}$$

This shows that the general condition for linear independence is violated for the specific matrix A. The coefficients are $c_1 = c_3 = \frac{1}{2}$ and $c_2 = -1$, and are thus different from zero.

Trace

The trace of a matrix $\text{tr}(\mathcal{A})$ is the sum of its diagonal elements:

$$\text{tr}(A) = \sum_{i=1}^{\min(n,p)} a_{ii}.$$

The trace of a scalar just equals the scalar itself. One obtains the trace in R by combining the functions `diag()` and `sum()`:

```
> A = matrix(1:12, nrow = 4, ncol = 3); A
     [,1] [,2] [,3]
[1,]    1    5    9
[2,]    2    6   10
[3,]    3    7   11
[4,]    4    8   12
> sum(diag(A))      # trace
[1] 18
```

The function `diag()` extracts the diagonal elements of a matrix, which are then summed by the function `sum()`.

Determinant

The formal definition of the determinant of a square matrix $A_{(p \times p)}$ is

$$\det(A) = \sum (-1)^{\phi_p(\tau_1,...,\tau_p)} a_{1\,\tau_1} \ldots a_{p\,\tau_p}, \tag{2.2}$$

where $\phi_p(\tau_1, \ldots, \tau_p) = n_1 + \cdots + n_p$ and n_k represents the number of integers in the subsequence $\tau_{k+1}, \ldots, \tau_p$ that are smaller than τ_k. For a square matrix $A_{(2 \times 2)}$ of dimension two, (2.2) reduces to

$$\det(A_{(2 \times 2)}) = a_{11}a_{22} - a_{12}a_{21}.$$

The determinant is often useful for checking whether matrices are singular or regular. If the determinant is equal to 0, then the matrix is singular. Singular matrices can not be inverted, which limits some computations. In R the determinant is computed by the function `det()`:

```
> A = matrix(1:9, nrow = 3, ncol = 3); A
     [,1] [,2] [,3]
[1,]    1    2    3
[2,]    4    5    6
[3,]    7    8    9
> det(A) # determinant
[1] 0
```

Thus, A is singular.

Transpose

A matrix $A_{(n \times p)}$ has a transpose $A^{\top}_{(n \times p)}$, which is obtained by reordering the elements of the original matrix. Formally, the transpose of $A_{(n \times p)}$ is

$$A^{\top}_{(n \times p)} = (a_{i\,j})^{\top} = (a_{j\,i}).$$

The resulting matrix has p rows and n columns. One has that

$$(A^{\top})^{\top} = A,$$
$$(AB)^{\top} = B^{\top} A^{\top}.$$

R provides the function `t()` which returns the transpose of a matrix:

```
> A = matrix(1:9, nrow = 3, ncol = 3, byrow = TRUE); A
     [,1] [,2] [,3]
[1,]    1    2    3
[2,]    4    5    6
[3,]    7    8    9
> t(A)   # transpose
     [,1] [,2] [,3]
[1,]    1    4    7
```

```
[2,]   2    5    8
[3,]   3    6    9
```

When creating a matrix with a constructor `matrix`, its transpose can be created by setting the argument `byrow` to `FALSE`.

A good overview of special matrices and vectors is provided by Table 2.1 in Härdle and Simar (2015), Chap. 2. The same notations are used in this book.

Conjugate transpose

Every matrix $\mathcal{A}_{(n \times p)}$ has a conjugate transpose $\mathcal{A}^C_{p \times n}$. The elements of $\mathcal{A}$ can be complex numbers. If a matrix entry $a_{ij} = \alpha + \beta i$ is a complex number with real numbers α, β and imaginary unit $i^2 = -1$, then its conjugate is $a_{ij}^C = \alpha - \beta i$. The same holds in the other direction: if $a_{ij} = \alpha - \beta i$, the conjugate is $a_{ij}^C = \alpha + \beta i$. Therefore the conjugate transpose is

$$\mathcal{A}^C = \begin{pmatrix} a_{11}^C & a_{21}^C & \dots & a_{n1}^C \\ a_{12}^C & a_{22}^C & \dots & a_{n2}^C \\ \vdots & \vdots & \ddots & \vdots \\ a_{p1}^C & \dots & \dots & a_{np}^C \end{pmatrix}. \tag{2.3}$$

The function `Conj()` yields the conjugates of the elements. One can combine the functions `Conj()` and `t()` to get the conjugate transpose of a matrix. For $\mathcal{A} = \begin{pmatrix} 1+0.5 \cdot i & 1 & 1 \\ 1 & 1 & 1-0.5 \cdot i \end{pmatrix}$, the conjugate transpose is computed in R as follows:

```
> a  = c(1 + 0.5i, 1, 1, 1, 1, 1 - 0.5i)            # matrix entries
> A  = matrix(a, nrow = 2, ncol = 3, byrow = TRUE)  # complex matrix
> A
        [,1]  [,2]    [,3]
[1,] 1+0.5i 1+0i 1+0.0i
[2,] 1+0.0i 1+0i 1-0.5i
> AC = Conj(t(A))                                   # conjugate
> AC                                                # transpose
        [,1]    [,2]
[1,] 1-0.5i 1+0.0i
[2,] 1+0.0i 1+0.0i
[3,] 1+0.0i 1+0.5i
```

For a matrix with only real values, the conjugate transpose $\mathcal{A}^C$ is equal to the normal transpose $\mathcal{A}^\top$.

2.1.2 Matrix Operations

There are four fundamental operations in arithmetic: addition, subtraction, multiplication and division. In matrix algebra, there exist analogous operations: matrix addition, matrix subtraction, matrix multiplication and 'division'.

Basic operations

For matrices $\mathcal{A}_{(n \times p)}$ and $\mathcal{B}_{(n \times p)}$ of the same dimensions, matrix addition and subtraction work elementwise as follows:

$$\mathcal{A} + \mathcal{B} = (a_{ij} + b_{ij}),$$

$$\mathcal{A} - \mathcal{B} = (a_{ij} - b_{ij}).$$

These operations can be applied in R as shown below.

```
> A = matrix(3:11, nrow = 3, ncol = 3, byrow = TRUE); A
     [,1] [,2] [,3]
[1,]    3    4    5
[2,]    6    7    8
[3,]    9   10   11
> B = matrix(-3:-11, nrow = 3, ncol = 3, byrow = TRUE); B
     [,1] [,2] [,3]
[1,]   -3   -4   -5
[2,]   -6   -7   -8
[3,]   -9  -10  -11
> A + B
     [,1] [,2] [,3]
[1,]    0    0    0
[2,]    0    0    0
[3,]    0    0    0
```

R reports an error if one tries to add or subtract matrices with different dimensions. The elementary operations, including addition, subtraction, multiplication and division can also be used with a scalar and a matrix in R, and are applied to each entry of the matrix. An example is the modulo operation

```
> A = matrix(1:9, nrow = 3, ncol = 3, byrow = TRUE); A
     [,1] [,2] [,3]
[1,]    1    2    3
[2,]    4    5    6
[3,]    7    8    9
> A %% 2   # modulo operation
     [,1] [,2] [,3]
[1,]    1    0    1
[2,]    0    1    0
[3,]    1    0    1
```

If one uses in R the elementary operations, including addition +, subtraction −, multiplication *, or division/between two matrices, they are all interpreted in R as elementwise operations.

Matrix multiplication returns the matrix product of the matrices $\mathcal{A}_{(n \times p)}$ and $\mathcal{B}_{(p \times m)}$, which is

$$A_{(n \times p)} \cdot B_{(p \times m)} = C_{(n \times m)} = \begin{pmatrix} \sum_{i=1}^{p} a_{1i}b_{i1} & \sum_{i=1}^{p} a_{1i}b_{i2} & \cdots & \sum_{i=1}^{p} a_{1i}b_{im} \\ \sum_{i=1}^{p} a_{2i}b_{i1} & \sum_{i=1}^{p} a_{2i}b_{i2} & \cdots & \sum_{i=1}^{p} a_{2i}b_{im} \\ \vdots & \vdots & \ddots & \vdots \\ \sum_{i=1}^{p} a_{ni}b_{i1} & & \cdots & \cdots & \sum_{i=1}^{p} a_{ni}b_{im} \end{pmatrix}.$$

In R, one uses the operator %*% between two objects for matrix multiplication. The objects have to be of class `vector` or `matrix`.

```
> A = matrix(3:11, nrow = 3, ncol = 3, byrow = TRUE); A
     [,1] [,2] [,3]
[1,]    3    4    5
[2,]    6    7    8
[3,]    9   10   11
> B = matrix(-3:-11, nrow = 3, ncol = 3, byrow = TRUE); B
     [,1] [,2] [,3]
[1,]   -3   -4   -5
[2,]   -6   -7   -8
[3,]   -9  -10  -11
> A %*% B # matrix multiplication
      [,1] [,2] [,3]
[1,]   -78  -90 -102
[2,]  -132 -153 -174
[3,]  -186 -216 -246
```

The number of columns of A has to equal the number of rows of B.

Inverse

The division operation for square matrices is done by inverting a matrix. The inverse A^{-1} of a square matrix $A_{(p \times p)}$ exists if $\det(A) \neq 0$:

$$A^{-1}A = AA^{-1} = I_p. \tag{2.4}$$

The inverse of $A = (a_{ij})$ can be calculated by

$$A^{-1} = \frac{W}{\det(A)},$$

where $W = (w_{ij})$ is the adjoint matrix of A. The elements of W are

$$(w_{ji}) = (-1)^{i+j} \det \begin{pmatrix} a_{11} & \cdots & a_{1(j-1)} & a_{1(j+1)} & \cdots & a_{1p} \\ \vdots & \vdots & \vdots & \vdots & \vdots & \vdots \\ a_{(i-1)1} & \cdots & a_{(i-1)(j-1)} & a_{(i-1)(j+1)} & \cdots & a_{(i-1)p} \\ a_{(i+1)1} & \cdots & a_{(i+1)(j-1)} & a_{(i+1)(j+1)} & \cdots & a_{(i+1)p} \\ \vdots & \vdots & \vdots & \vdots & \vdots & \vdots \\ a_{p1} & \cdots & a_{p(j-1)} & a_{p(j+1)} & \cdots & a_{pp} \end{pmatrix},$$

which are the cofactors of $\mathcal{A}$. To compute the cofactors w_{ji}, one deletes column j and row i of $\mathcal{A}$, then computes the determinant for that reduced matrix, and then multiplies by 1 if $j + i$ is even or by -1 if it is odd. This computation is only feasible for small matrices.

Using the above definition, one can determine the inverse of a square matrix by solving the system of linear equations (see Sect. 2.4.1) in (2.4) by employing the function solve(A, b). In R this function can be used to solve a general system of linear equations $Ax = b$. If one does not specify the right side b of the system of equations, the solve() function computes the inverse of the square matrix $\mathcal{A}$. The following code computes the inverse of the square matrix $\mathcal{A} = \left(\begin{smallmatrix} 1 & 2 & 5 \\ 3 & 9 & 2 \\ 2 & 2 & 2 \end{smallmatrix}\right)$.

```
> A = matrix(c(1, 2, 5, 3, 9, 2, 2, 2, 2),   # all elements
+     nrow = 3, ncol = 3, byrow = TRUE); A    # matrix dimensions
      [,1] [,2] [,3]
[1,]    1    2    5
[2,]    3    9    2
[3,]    2    2    2
> solve(A)                                    # inverse of A
       [,1]  [,2]  [,3]
[1,] -0.28 -0.12  0.82
[2,]  0.04  0.16 -0.26
[3,]  0.24 -0.04 -0.06
> A %*% solve(A)                              # check (2.4)
              [,1]          [,2]          [,3]
[1,]  1.000000e+00  5.551115e-17 -1.110223e-16
[2,] -5.551115e-17  1.000000e+00  8.326673e-17
[3,] -5.551115e-17  1.804112e-16  1.000000e+00
```

For diagonal matrices, the inverse $\mathcal{A}^{-1} = (a_{ii}^{-1})$ if $a_{ii} \neq 0$ for all i.

Generalised inverse

In practice, we are often confronted with singular matrices, whose determinant is equal to zero. In this situation, the inverse can be given by a generalised inverse $\mathcal{A}^-$ satisfying

$$\mathcal{A}\mathcal{A}^-\mathcal{A} = \mathcal{A}. \qquad (2.5)$$

Consider the singular matrix $\mathcal{A} = \left(\begin{smallmatrix} 1 & 0 \\ 0 & 0 \end{smallmatrix}\right)$. Its inverse $\mathcal{A}^-$ must satisfy (2.5),

$$\begin{pmatrix} 1 & 0 \\ 0 & 0 \end{pmatrix} \mathcal{A}^- \begin{pmatrix} 1 & 0 \\ 0 & 0 \end{pmatrix} = \begin{pmatrix} 1 & 0 \\ 0 & 0 \end{pmatrix}. \qquad (2.6)$$

There are sometimes several $\mathcal{A}^-$ which satisfy (2.5). The Moore–Penrose generalised inverse (hereafter, just 'generalised inverse') is the most common type and was developed by Moore (1920) and Penrose (1955). It is used to compute the 'best fit' solution to a system of linear equations that does not have a unique solution. Another approach is to find the minimum (Euclidean) norm (see Sect. 2.1.5) solution to a system of linear equations with several solutions. The Moore–Penrose generalised inverse is defined and unique for all matrices with real or complex entries. It can be computed using the singular value decomposition, see Press (1992).

In R the generalised inverse of a matrix defined in (2.5) can be computed with the function ginv() from the MASS package. With ginv() one obtains the generalised inverse of the matrix $\mathcal{A} = \left(\begin{smallmatrix} 1 & 0 \\ 0 & 0 \end{smallmatrix}\right)$, which is equal to $\mathcal{A}^- = \left(\begin{smallmatrix} 1 & 0 \\ 0 & 0 \end{smallmatrix}\right)$.

```
> require(MASS)
> A = matrix(c(1, 0, 0, 0),
+    ncol = 2, nrow = 2); A        # matrix from (2.6)
     [,1] [,2]
[1,]   1    0
[2,]   0    0
> ginv(A)                          # generalised inverse
     [,1] [,2]
[1,]   1    0
[2,]   0    0
```

The ginv() function can also be used for non-square matrices, like $\mathcal{A} = \left(\begin{smallmatrix} 1 & 2 & 3 \\ 11 & 12 & 13 \end{smallmatrix}\right)$.

```
> require(MASS)
> A = matrix(c(1, 2, 3, 11, 12, 13),
+    nrow = 2, ncol = 3, byrow = TRUE); A # non-square matrix
     [,1] [,2] [,3]
[1,]    1    2    3
[2,]   11   12   13
> A.ginv = ginv(A)                          # generalised inverse
             [,1]         [,2]
[1,] -0.63333333  0.13333333
[2,] -0.03333333  0.03333333
[3,]  0.56666667 -0.06666667
```

The condition (2.5) can be verified in R by the following code:

```
> A %*% A.ginv %*% A                  # first condition
     [,1] [,2] [,3]
[1,]    1    2    3
[2,]   11   12   13
```

This code shows that the solution for A^- fulfills the condition $\mathcal{A}\mathcal{A}^-\mathcal{A} = \mathcal{A}$.

2.1.3 Eigenvalues and Eigenvectors

For a given basis of a vector space, a matrix $\mathcal{A}_{(p \times p)}$ can represent a linear function of a p-dimensional vector space to itself. If this function is applied to a nonzero vector and maps that vector to a multiple of itself, that vector is called an eigenvector γ and the multiple is called the corresponding eigenvalue λ. Formally this can be written as

$$\mathcal{A}\gamma = \lambda\gamma.$$

In Chap. 8, the theory of eigenvectors, eigenvalues and spectral decomposition, presented in Sect. 2.1.4, becomes important in order to understand the *principal components analysis*, the *factor analysis* and other methods of dimension reduction. Definition 2.1 provides a formal description of an eigenvalue and the corresponding eigenvector.

Definition 2.1 Let V be any real vector space and $L : V \to V$ be a linear transformation. Then a nonzero vector $\gamma \in V$ is called an eigenvector if, and only if, there exists a scalar $\lambda \in \mathbb{R}$ such that $L(\gamma) = \mathcal{A}\gamma = \lambda\gamma$. Therefore

$$\lambda \text{ is eigenvalue of } \mathcal{A} \Leftrightarrow \det(\mathcal{A} - \lambda\mathcal{I}) = 0.$$

If we consider λ as an unknown variable, then $\det(\mathcal{A} - \lambda\mathcal{I})$ is a polynomial of degree n in λ, and is called the characteristic polynomial of $\mathcal{A}$. Its coefficients can be computed from the matrix entries of $\mathcal{A}$. Its roots $\lambda_1, \ldots, \lambda_p$, which might be complex, are the eigenvalues of $\mathcal{A}$. The eigenvalue matrix Λ is a diagonal matrix with elements $\lambda_1, \ldots, \lambda_p$. Vectors $\gamma_1, \ldots, \gamma_p$ are the eigenvectors, that correspond to eigenvalues $\lambda_1, \ldots, \lambda_p$. The eigenvector matrix $\mathcal{P}$ has columns $\gamma_1, \ldots, \gamma_p$. In the following example, the computation of the eigenvalues of a matrix of dimension 3 is shown.

Example 2.1 Consider the matrix $\mathcal{A} = \begin{pmatrix} 2 & 0 & 1 \\ 0 & 3 & 1 \\ 0 & 6 & 2 \end{pmatrix}$ and the matrix $\mathcal{D} = \mathcal{A} - \lambda\mathcal{I}$.

In order to obtain the eigenvalues of $\mathcal{A}$ one has to solve for the roots λ of the polynomial $\det(\mathcal{D}) = 0$. For a three-dimensional matrix this looks like

$$\det(\mathcal{D}) = c_0 + c_1\lambda + c_2\lambda^2 + c_3\lambda^3,$$
$$c_0 = \det(\mathcal{A}),$$
$$c_1 = \sum_{1 \le i \ne j \le 3} a_{ii}a_{jj} - \sum_{1 \le i \ne j \le 3} a_{ij}a_{ji},$$
$$c_2 = \sum_{1 \le i \le 3} a_{ii},$$
$$c_3 = -1.$$

Thus, the characteristic polynomial of $\mathcal{A}$ is $-\lambda^3 + 7\lambda^2 + 10\lambda$. In this case, $\mathcal{A}$ is singular: the intercept of the polynomial is equal to zero. Therefore, one eigenvalue is 0 and the other two are 2 and 5.

In R the eigenvalues and eigenvectors of a matrix $\mathcal{A}$ can be calculated using the function eigen().

```
> A = matrix(c(2, 0, 1, 0, 3, 1, 0, 6, 2),  # matrix A
+    nrow = 3, ncol = 3, byrow = TRUE); A
        [,1] [,2] [,3]
[1,]     2    0    1
[2,]     0    3    1
[3,]     0    6    2
> Eigen = eigen(A)                            # eigenvectors and -values
```

```
> Eigen$values                             # eigenvalues
[1] 5 2 0
> Eigen$vectors                            # eigenvector matrix
           [,1] [,2]        [,3]
[1,] 0.2857143    1 -0.4285714
[2,] 0.4285714    0 -0.2857143
[3,] 0.8571429    0  0.8571429
```

Let $\gamma_2 = (1, 0, 0)^\top$ be the second column of the eigenvector matrix $\mathcal{P} = (\gamma_1, \gamma_2, \gamma_3)$.
Then it can be seen that

$$\mathcal{A}\gamma_2 = 2\gamma_2.$$

This means that γ_2 is the eigenvector corresponding to the eigenvalue $\lambda_2 = 2$ of $\mathcal{A}$.

After the eigenvalues of a matrix are found, it is easy to compute its trace or determinant. The sum of all its eigenvalues is its trace: $\text{tr}(\mathcal{A}) = \sum_{i=1}^{p} \lambda_i$. The product of its eigenvalues is its determinant: $\det(\mathcal{A}) = \prod_{i=1}^{p} \lambda_i$.

2.1.4 Spectral Decomposition

The spectral decomposition of a matrix is a representation of that matrix in terms of its eigenvalues and eigenvectors.

Theorem 2.1 *Let $A_{(p \times p)}$ be a matrix with real entries and let Λ be its eigenvalue matrix and $\mathcal{P}$ the corresponding eigenvector matrix. Then*

$$\mathcal{A} = \mathcal{P}\Lambda\mathcal{P}^{-1}. \tag{2.7}$$

In R, one can use the function `eigen()` to compute eigenvalues and eigenvectors. The eigenvalues are in the field named `values` and are sorted in decreasing order (see the example above). Using the output of the function `eigen()`, the linear independence of the eigenvectors can be checked for the above example by computing the rank of the matrix $\mathcal{P}$:

```
> A = matrix(c(2, 0, 1, 0, 3, 1, 0, 6, 2),  # matrix A
+    nrow = 3, ncol = 3, byrow = TRUE); A
     [,1] [,2] [,3]
[1,]   2    0    1
[2,]   0    3    1
[3,]   0    6    2
> Eigen = eigen(A)                          # eigenvectors and -values
> P     = eigen(A)$vectors                  # eigenvector matrix
> L     = diag(eigen(A)$values)             # eigenvalue matrix
> qr(P)$rank                                # rank of P
[1] 3
> P %*% L %*% solve(P)                       # spectral decomposition
        [,1]              [,2] [,3]
[1,]   2 4.440892e-16        1
[2,]   0 3.000000e+00        1
[3,]   0 6.000000e+00        2
```

From this computation, it can be seen that $\mathcal{P}$ has full rank. The diagonal matrix can be obtained by extracting the eigenvalues from the output of the function `eigen()`. It is possible to decompose the matrix $\mathcal{A}$ by (2.7) in R. The difference between $\mathcal{A}$ and the result from the spectral decomposition in R is negligibly small.

2.1.5 Norm

There are two types of frequently used norms: the vector norm and the matrix norm. The vector norm, which appears frequently in matrix algebra and numerical computation, will be introduced first. An extension of the vector norm is the matrix norm.

Definition 2.2 Let V be a vector space and b be a scalar, both lying either in $\mathbb{R}^n$ or $\mathbb{C}^n$. Consider the vectors $x, y \in V$. Then a norm is a mapping $\| \cdot \| : V \to \mathbb{R}_0^+$ with the following properties:

1. $\|bx\| = |b|\|x\|$,
2. $\|x + y\| \leq \|x\| + \|y\|$,
3. $\|x\| \geq 0$, where $\|x\| = 0$ if and only if $x = 0$.

Let $x = (x_1, \ldots, x_n)^\top \in \mathbb{R}^n$, $k \geq 1$ and $k \in \mathbb{R}$. Then a general norm is the L_k norm, which can be represented as follows,

$$\|x\|_k = \left(\sum_{i=1}^{n} |x_i|^k \right)^{1/k}.$$

There are several special norms, depending on the value of k, some are listed below.

$$\text{Manhattan norm}: \|x\|_1 = \sum_{i=1}^{n} |x_i|, \tag{2.8}$$

$$\text{Euclidean norm}: \|x\|_2 = \left(\sum_{i=1}^{n} |x_i|^2 \right)^{1/2} = \sqrt{x^\top x}, \tag{2.9}$$

$$\text{infinity norm}: \|x\|_\infty = \max\{|x_i|_{i=1}^n\}, \tag{2.10}$$

$$\text{Frobenius norm}: \|x\|_F = \sqrt{x^\top x}. \tag{2.11}$$

The most frequently used norms are the Manhattan and Euclidean norms. For vector norms, the Euclidean and Frobenius norm coincide. The infinity norm selects the maximum absolute value of the elements of x and the maximum norm just the maximum value.

In R the function `norm()` can return the norms from (2.8) to (2.11). The argument `type` specifies which norm is returned.

```
> x = matrix(c(2, 1, 2), nrow = 3, ncol = 1) # vector x
> norm(x, type = c("O"))                       # Manhattan norm
[1] 5
> norm(x, type = c("2"))                       # Euclidean norm
[1] 3
> norm(x, type = c("I"))                       # infinity norm
[1] 2
> norm(x, type = c("F"))                       # Frobenius norm
[1] 3
```

The object x has to be of class `matrix` in R to compute all norms.

Definition 2.3 Let $U^{n \times p}$ be a set of $(n \times p)$ matrices and a be a scalar, which are either real or complex. $U^{n \times p}$ is a vector space equipped with matrix addition and scalar multiplication. Let $\mathcal{A}, \mathcal{B} \in U^{n \times p}$. Then a matrix norm is a mapping $\| \cdot \|$: $U^{n \times p} \to \mathbb{R}_0^+$ with the following properties:

1. $\|a\mathcal{A}\| = |a|\|\mathcal{A}\|$,
2. $\|\mathcal{A} + \mathcal{B}\| \le \|\mathcal{A}\| + \|\mathcal{B}\|$,
3. $\|\mathcal{A}\| \ge 0$, where $\|\mathcal{A}\| = 0$ if and only if $\mathcal{A} = 0$.

In R, the function `norm()` can be applied to vectors and matrices in the same fashion. The one norm, the infinity norm, the Frobenius norm, the maximum norm and the spectral norm for matrices are represented by

$$\text{one norm} : \|\mathcal{A}\|_1 = \max_{1 \le j \le p} \sum_{i=1}^{n} |a_{ij}|,$$
$$\text{spectral/Euclidean norm} : \|\mathcal{A}\|_2 = \sqrt{\lambda_{max}(\mathcal{A}^C \mathcal{A})},$$
$$\text{infinity norm} : \|\mathcal{A}\|_\infty = \max_{1 \le i \le n} \sum_{j=1}^{p} |a_{ij}|,$$
$$\text{Frobenius norm} : \|\mathcal{A}\|_F = \sqrt{\sum_{i=1}^{n} \sum_{j=1}^{p} |a_{ij}|^2},$$

where $\mathcal{A}^C$ is the conjugate matrix of $\mathcal{A}$. The next code shows how to compute these five norms with the function `norm()` for the matrix $\mathcal{A} = \left(\begin{smallmatrix} 1 & 2 \\ 3 & 4 \end{smallmatrix}\right)$.

```
> A = matrix(c(1, 2, 3, 4),
+   ncol = 2, nrow = 2, byrow = TRUE)    # matrix A
> norm(A, type = c("O"))                  # one norm
[1] 6                                     # maximum of column sums
> norm(A, type = c("2"))                  # Euclidean norm
[1] 5.464986
> norm(A, type = c("I"))                  # infinity norm
[1] 7                                     # maximum of row sums
> norm(A, type = c("F"))                  # Frobenius norm
[1] 5.477226
```

Note that the Frobenius norm returns the square root of the trace of the matrix product of the matrix with its conjugate transpose, $\sqrt{\text{tr}(\mathcal{A}^C \mathcal{A})}$. The spectral norm or Euclidean norm returns the square root of the maximum eigenvalue of $\mathcal{A}^C \mathcal{A}$.

2.2 Numerical Integration

This section discusses numerical methods in R for integrating a function. Some integrals cannot be computed analytically and numerical methods should be used.

2.2.1 Integration of Functions of One Variable

Not every function $f \in C[a, b]$ has an indefinite integral with an analytical representation. Therefore, it is not always possible to analytically compute the area under a curve. An important example is

$$\int \exp(-x^2)dx. \tag{2.12}$$

There exists no analytical, closed-form representation of (2.12). Therefore, the corresponding definite integral has to be computed numerically using numerical integration, also called 'quadrature'. The basic idea behind numerical integration lies in approximating the function by a polynomial and subsequently integrating it using the Newton–Cotes rule.

Newton–Cotes rule

If a function $f \in C[a, b]$ and nodes $a = x_0 < x_1 < \cdots < x_n = b$ are given, one looks for a polynomial $p_n(x) = \sum_{j=1}^{n} c_j x^{j-1} \in P_n$, where P_n are basis polynomials, satisfying the condition

$$p_n(x_k) = f(x_k), \quad \text{for all } k \in \{0, \ldots, n\}. \tag{2.13}$$

To construct a polynomial that satisfies this condition, the following basis polynomials are used:

$$L_k(x) = \prod_{i=0, i \neq k}^{n} \frac{x - x_i}{x_k - x_i}.$$

This leads to the so-called *Lagrange polynomial*, which satisfies the condition in (2.13) (assuming $\frac{0}{0} = 1$).

$$p_n(x) = \sum_{k=0}^{n} f(x_k) L_k(x). \tag{2.14}$$

Let $I(f) = \int_a^b f(x)dx$ be the exact integration operator applied to a function $f \in C[a, b]$. Then define $I_n(f)$ as the approximation of $I(f)$ using (2.14) as an approximation for f:

$$I_n(f) = \int_a^b p_n(x)\,dx,$$

which can be restated using weights for the different values of f:

$$I_n(f) = (b - a) \sum_{k=0}^{n} f(x_k)\alpha_k,\tag{2.15}$$

$$\text{with weights} \quad \alpha_k = \frac{1}{b-a} \int_a^b L_k(x)\,dx.$$

By construction, (2.15) is exact for every $f \in P_n$. Suppose the nodes x_k are equidistant in $[a, b]$, i.e., $x_k = a + kh$, where $h = (b - a)n^{-1}$. Then (2.15) is the (closed) *Newton–Cotes rule*. The weights α_k can be explicitly computed up to $n = 7$. Starting from $n = 8$, negative weights occur and the Newton–Cotes rule can no longer be applied. The *trapezoidal rule* is an example of the Newton–Cotes rule.

Example 2.2 For $n = 1$ and $I(f) = \int_a^b f(x)\,dx$, the nodes are given as follows: $x_0 = a$, $x_1 = b$. The weights can be computed explicitly by transforming the integral using two substitutions:

$$\alpha_k = \frac{1}{b-a} \int_a^b L_k(x)\,dx = \int_0^1 \prod_{i=0,i\neq k}^{n} \frac{t - t_i}{t_k - t_i}\,dt = \frac{1}{n} \int_0^n \prod_{i=0,i\neq k}^{n} \frac{s - i}{k - i}\,ds.$$

Then the weights for $n = 1$ are $\alpha_0 = \frac{1}{2}$ and $\alpha_1 = \frac{1}{2}$. So the Newton–Cotes rule $I_1(f)$ is given by the formula for the area of a trapezoid:

$$I_1(f) = (b - a) \left\{ \frac{f(a) + f(b)}{2} \right\}.$$

In R, the trapezoidal rule is implemented within the package caTools. There the function trapz (x, y) is used with a sorted vector x that contains the x-axis values and a vector y with the corresponding y-axis values. This function uses a summed version of the trapezoidal rule, where $[a, b]$ is split into n equidistant intervals. For all $k = \{1, \ldots, n\}$, the integral $I_k(f)$ is computed according to the trapezoidal rule: this is the so-called *extended trapezoidal rule*.

$$I_k(f) = \frac{b - a}{2n}\left[f\{a + (k - 1)n^{-1}(b - a)\} + f\{a + kn^{-1}(b - a)\}\right].$$

Therefore, the whole integral $I(f)$ is approximated by

$$I(f) \approx \sum_{k=1}^{n} I_k(f) = \sum_{k=1}^{n} \frac{b - a}{2n}\left[f\{a + (k - 1)n^{-1}(b - a)\} + f\{a + kn^{-1}(b - a)\}\right].$$

For example, consider the integral of the cosine function on $[-\frac{\pi}{2}, \frac{\pi}{2}]$ and split the interval into 10 subintervals, where the trapezoidal rule is applied:

```
> require(caTools)
> x        = (-5:5) * (pi / 2) / 5     # set subintervals
> intcos = trapz(x, cos(x)); intcos    # integration
[1] 1.983524
> abs(intcos - 2)                      # absolute error
[1] 0.01647646
```

The integral of the cosine function on $[-\frac{\pi}{2}, \frac{\pi}{2}]$ is supposed to be exactly 2, so the absolute error is almost 0.02. It can be shown that the error of the trapezoidal rule is of order $\mathcal{O}(h^3 \| f'' \|^2)$ with an unknown second derivative of the function. If a Newton–Cotes rule with more nodes is used, the integrand will be approximated by a polynomial of higher order. Therefore, the error could diminish if the integrand is very smooth, so that it can be approximated well by a polynomial.

Gaussian quadrature

In R, the function `integrate()` uses an integration method that is based on Gaussian quadrature (the exact method is called the *Gauss–Kronrod quadrature*, see Press (1992) for further details). The Gaussian method uses non-predetermined nodes $x_1, \ldots, x_n$ to approximate the integral, so that polynomials of higher order can be integrated more precisely than using the Newton–Cotes rule. For n nodes, it uses a polynomial $p(x) = \sum_{j=1}^{2n} c_j x^{j-1}$ of order $2n - 1$.

Definition 2.4 A method of numerical integration for a function $f : [a, b] \to \mathbb{R}$ with the formula

$$I(f) = \int_a^b f(x)dx \approx \int_a^b w(x)p(x)dx \approx I_n^G(f) = \sum_{k=1}^n f(x_k)\alpha_k,$$

and n nodes $x_1, \ldots, x_n$ is called Gaussian quadrature if an arbitrary weighting function $w(x)$ and a polynomial $p(x)$ of degree $2n - 1$ exactly approximate $f(x)$, such that $f(x) = w(x)p(x)$.

Consider the simplest case, where $w(x) = 1$. Then the method of undetermined coefficients leads to the following nonlinear system of equations:

$$\frac{b^j - a^j}{j!} = \sum_{k=1}^n \alpha_k x_k^{j-1}, \quad \text{for } j = 1, \ldots, 2n.$$

A total of $2n$ equations are used to find the nodes $x_1, \ldots, x_n$ and the coefficients $\alpha_1, \ldots, \alpha_n$.

Consider the special case with two nodes (x_1, x_2) and two weights (α_1, α_2). The particular polynomial $p(x)$ is of order $2 \cdot n - 1 = 3$, where the number of nodes is n. The integral is approximated by $\alpha_1 f(x_1) + \alpha_2 f(x_2)$ and it is assumed that $f(x) = p(x)$. Therefore the following two equations can be derived:

$$I_2^G(f) = c_1(\alpha_1 + \alpha_2) + c_2(\alpha_1 x_1 + \alpha_2 x_2) + c_3(\alpha_1 x_1^2 + \alpha_2 x_2^2) + c_4(\alpha_1 x_1^3 + \alpha_2 x_2^3),$$

$$I_2^G(f) = c_1(b - a) + c_2\left(\frac{b^2 - a^2}{2}\right) + c_3\left(\frac{b^3 - a^3}{3}\right) + c_4\left(\frac{b^4 - a^4}{4}\right).$$

All coefficients c_j of the polynomial are set equal to each other, because the coefficients are arbitrary. The system of four nonlinear equations is

$$b - a = \alpha_1 + \alpha_2;$$
$$1/2 \cdot (b^2 - a^2) = \alpha_1 x_1 + \alpha_2 x_2;$$
$$1/3 \cdot (b^3 - a^3) = \alpha_1 x_1^2 + \alpha_2 x_2^2;$$
$$1/4 \cdot (b^4 - a^4) = \alpha_1 x_1^3 + \alpha_2 x_2^3.$$

For simplicity, in most cases the interval $[-1, 1]$ is considered. It is possible to extend these results to the more general interval $[a, b]$. To apply the results for $[-1, 1]$ to the interval $[a, b]$, one uses

$$\int_a^b f(x)dx = \frac{b - a}{2} \int_{-1}^1 f\left(\frac{b - a}{2}x + \frac{a + b}{2}\right)dx.$$

For the special case $w(x) = 1$ and the interval $[-1, 1]$, the procedure is called *Gauss–Legendre quadrature*. The nodes are the roots of the *Legendre polynomials* $P_n(x) = \frac{1}{2^n \cdot n!} \frac{d^n}{dx^n}\{(x^2 - 1)^n\}$. The weights α_k can be calculated by

$$\alpha_k = \frac{2}{(1 - x_k^2)\{P_n'(x_k)\}^2}.$$

In the following example, we illustrate the process of numerical integration using the function `integrate()`. One can specify the following arguments: `f` (integrand), `a` (lower limit) and `b` (upper limit), `subdivisions` (number of subintervals) and arguments `rel.tol`, as well as `abs.tol` for the relative and absolute accuracy requested. Consider again the cosine function on the interval $[-\frac{\pi}{2}, \frac{\pi}{2}]$.

```
> require(stats)
> integrate(cos,        # integrand
+    lower = -pi / 2,  # lower integration limit
+    upper = pi / 2)   # upper integration limit
2 with absolute error < 2.2e-14
```

The output of the `integrate()` function delivers the computed value of the definite integral and an upper bound on the absolute error. In this example, the absolute error is smaller than $2.2 \cdot 10^{-14}$. Therefore, the `integrate()` function is much more accurate for the cosine function than the `trapz()` function used in a previous example.

2.2.2 Integration of Functions of Several Variables

Repeated quadrature method

Similar to numerical integration in the context of one variable, in the case of more variables, an integration can be expressed as follows:

$$\int_{a_1}^{b_1} \cdots \int_{a_p}^{b_p} f(x_1, \ldots, x_n)dx_1 \ldots dx_p \approx \sum_{i_1=1}^{n} \cdots \sum_{i_p=1}^{n} W_{i_1} \cdots W_{i_p} f(x_{i_1}, \ldots, x_{i_p}), \quad (2.16)$$

where $D_j = [a_j, b_j]$, $j \in \{1, \ldots, p\}$ is an integration region in $\mathbb{R}$ and $(x_{i_1}, \ldots, x_{i_p})$ is the p-dimensional point at the i-th dimension, where $i_j \in \{1, \ldots, n\}$, and W_{i_j} is the coefficient used as the weight. The problem with the repeated quadrature is that for (2.16) one needs to evaluate p^n terms, which may lead to computational difficulties.

Adaptive method

The adaptive method in the context of multiple integrals divides the integration region $D \in \mathbb{R}^p$ into subregions $S_j \in \mathbb{R}^p$. For each subregion S_j, specific rules are applied to approximate the integral. To improve the approximation, consider the error E_j for each subregion. If the overall error $\sum_j E_j$ is smaller than a predefined tolerance level, the algorithm stops. But if this condition is not met, the highest error is selected and the corresponding region is split into two subregions. Then the rules are applied again to each subregion until the tolerance level is met. For a more detailed description of this algorithm, see van Dooren and de Ridder (1976) and Stroud (1971).

Jarle Berntsen and Genz (1991) improved the reliability of the algorithm, including a strategy for the selection of subregions, error estimation and parallelisation of the computation.

Example 2.3 Integrate the function of two variables

$$\int_0^1 \int_0^1 x^2 y^3 dx dy. \quad (2.17)$$

Analytical computation of this integral yields $1/12$. The surface $z = x^2 y^3$ for the interval $[0, 1]^2$ is depicted in Fig. 2.1. For the computation of multiple integrals, the R package R2Cuba is used, which is introduced in Hahn (2013). It includes four different algorithms for multivariate integration, where the function cuhre uses the adaptive method.

```
> require(R2Cuba)
> integrand = function(arg){        # construct the integrand
+     x = arg[1]                     # function argument x
+     y = arg[2]                     # function argument y
+     (x^2) * (y^3)                  # function
+ }
> cuhre(integrand,                   # adaptive method
+     ncomp   = 1,                   # number of components
+     ndim    = 2,                   # dimension of the integral
```

```
+    lower   = rep(0, 2),          # lower integration bound
+    upper   = rep(1, 2),          # upper integration bound
+    rel.tol = 1e-3,               # relative tolerance level
+    abs.tol = 1e-12,              # absolute tolerance level
+    flags   = list(verbose = 1))  # controls output
Iteration 1:  65 integrand evaluations so far
[1] 0.0833333 +- 1.15907e-015   chisq 0 (0 df)
Iteration 2:  195 integrand evaluations so far
[1] 0.0833333 +- 1.04612e-015   chisq 0.10352e-05 (1 df)
integral: 0.08333333 (+-1.3e-15)
nregions: 2; number of evaluations: 195; probability: 0.00623341
```

The output shows that the adaptive algorithm carried out two iteration steps. Only two subregions have been used for the computation, which is stated by the output value `nregions`. The output value `neval` states that the number of evaluations is 195. To make a statement about the reliability of the process, consider the `probability` value. A probability of 0 for the χ^2 distribution (see Sect. 4.4.1) means that the null

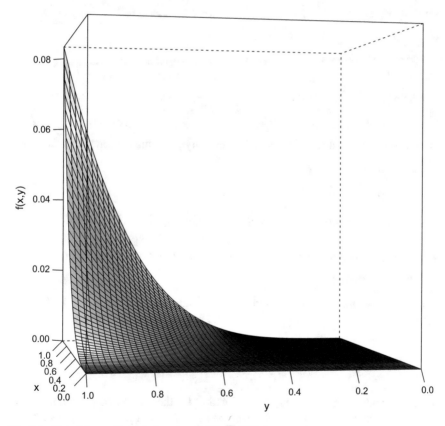

Fig. 2.1 Plot of the multivariate function (2.17). ⌀ BCS_Integrand

hypothesis can be rejected. The null hypothesis states that the absolute error estimate is not a reliable estimate of the true integration error. The approximation of integral I is 0.08333, which is close to the result of the analytical computation, $\frac{1}{12}$. For a more detailed discussion of the output, refer to Hahn (2013).

Example 2.4 Evaluate the integral with three variables

$$\int_0^1 \int_0^1 \int_0^1 \sin(x) \log(1 + 2y) \exp(3z) dx dy dz, \qquad (2.18)$$

```
> require(R2Cuba)
> integrand = function(arg){          # construct the integrand
+    x = arg[1]                        # function argument x
+    y = arg[2]                        # function argument y
+    z = arg[3]                        # function argument z
+    sin(x) * log(1 + 2 * y) * exp(3 * z)   # function
+ }
> cuhre(integrand,                     # adaptive method
+    ncomp   = 1,                       # number of components
+    ndim    = 3,                       # dimension of the integral
+    lower   = rep(0, 3),               # lower bound of interval
+    upper   = rep(1, 3),               # upper bound of interval
+    rel.tol = 1e-3,                    # relative tolerance level
+    abs.tol = 1e-12,                   # absolute tolerance level
+    flags   = list(verbose = 0))       # controls output
integral: 1.894854 (+-4.1e-07)
nregions: 2; number of evaluations:  381; probability:  0.04784836
```

For the function of three variables (2.18), an analytical computation yields the value:

$$\int_0^1 \sin(x) dx \int_0^1 \log(1 + 2y) dy \int_0^1 \exp(3z) dz$$

$$= \{1 - \cos(1)\} \frac{1}{2}[3\{\log(3) - 1\} + 1] \frac{1}{3}\{\exp(3) - 1\} = 1.89485.$$

The value provided by the adaptive method is very close to the exact value.

Monte Carlo method

For a multiple integral I of the function of p variables $f(x_1, \ldots, x_p)$ with lower bounds $a_1, \ldots, a_p$ and upper bounds $b_1, \ldots, b_p$, the integral is given by

$$I(f) = \int_{a_1}^{b_1} \ldots \int_{a_p}^{b_p} f(x_1, \ldots, x_p) dx_1 \ldots dx_p = \int \cdots \int_D f(x) dx,$$

where x stands for a vector $(x_1, \ldots, x_p)$ and D for the integration region. Let X be a random vector (see Chap. 6), with each component X_j of X uniformly distributed (Sect. 4.2) in $[a_j, b_j]$. Then the algorithm of Monte Carlo multiple integration can be described as follows. In the first step, n points of dimension p are randomly drawn from the region D, such that

$$(x_{1_1}, \ldots, x_{1_p}), \ldots, (x_{n_1}, \ldots, x_{n_p}).$$

In the second step, the p-dimensional volume is estimated by $V = \prod_{j=1}^{p}(b_j - a_j)$ and the integrand f is evaluated for all n points. In the third step, the integral I can be estimated using a sample moment function,

$$I(f) \approx \hat{I}(f) = n^{-1}V \sum_{i=1}^{n} f(x_{i_1}, \ldots, x_{i_p}).$$

The absolute error ϵ can be approximated as follows:

$$\epsilon = |(I(f) - \hat{I}(f)| \approx n^{-1/2}\{VI(f^2) - I^2(f)\}.$$

The Monte Carlo method is applied to example (2.17) via the function vegas.

```
> require(R2Cuba)
> integrand = function(arg){            # construct the integrand
+     x = arg[1]                         # function argument x
+     y = arg[2]                         # function argument y
+     (x^2) * (y^3)                      # function
+ }
> vegas(integrand,                       # Monte Carlo method
+     ncomp    = 1,                      # number of components
+     ndim     = 2,                      # dimension of the integral
+     lower    = rep(0, 2),              # lower integration bound
+     upper    = rep(1, 2),              # upper integration bound
+     rel.tol  = 1e-3,                   # relative tolerance level
+     abs.tol  = 1e-12,                  # absolute tolerance level
+     flags    = list(verbose = 0))      # controls output
integral: 0.08329357 (+-7.5e-05)
number of evaluations:  17500; probability:  0.1201993
```

The outputs of the functions vegas and cuhre are almost identical. Additional output information can be obtained by setting the argument verbose to one. Then the output shows that the Monte Carlo algorithm executed 7 iterations and 17 500 evaluations of the integrand. The approximation of integral I is 0.0832, which is close to the exact value $\frac{1}{12}$. For the function (2.18) the Monte Carlo algorithm looks as follows:

```
> require(R2Cuba)
> integrand = function(arg){            # construct the integrand
+     x = arg[1]                         # function argument x
+     y = arg[2]                         # function argument y
+     z = arg[3]                         # function argument z
+     sin(x) * log(1 + 2 * y) * exp(3 * z) # function
+ }
> vegas(integrand,                       # Monte Carlo method
+     ncomp    = 1,                      # number of components
+     ndim     = 3,                      # dimension of the integral
+     lower    = rep(0, 3),              # lower integration bound
+     upper    = rep(1, 3),              # upper integration bound
+     rel.tol  = 1e-3,                   # relative tolerance level
+     abs.tol  = 1e-12,                  # absolute tolerance level
+     flags    = list(verbose = 0))      # controls output
```

```
integral: 1.894488 (+-0.0016)
number of evaluations:   13500; probability:   0.1099108
```

The performance of the adaptive method is again superior to that of the Monte Carlo method, which gives 1.894488 as the value of the integral.

2.3 Differentiation

The analytical computation of derivatives may be impossible if the function is only given indirectly (for example by an algorithm) and can be evaluated only point-wise. Therefore, it is necessary to use numerical methods. Before presenting some numerical methods for differentiation, first it will be shown how to compute analytically the derivative in R.

2.3.1 Analytical Differentiation

To calculate the derivative of a one variable function in R, the function D(expr, name) is used. For expr the function is inserted (as an object of mode expression) and name identifies the variable with respect to which the derivative will be computed. Consider the following example:

```
> f = expression(3 * x^3 + x^2)
> D(f, "x")
3 * (3 * x^2) + 2 * x
```

The function D() returns an argument of type call (see help(call) for further information) and one can therefore recursively compute higher order derivatives. For example, consider the second derivative of $3x^3 + x^2$.

```
> D(D(f, "x"), "x")
3 * (3 * (2 * x)) + 2
```

To compute higher order derivatives, it can be useful to define a recursive function.

```
> DD = function(expr, name, order = 1){
+    if(order < 1) stop("'order' must be >= 1")# warning message
+    if(order == 1) D(expr, name)             # first derivative
+    else DD(D(expr, name), name, order - 1)  # 1st derivative of DD
+ }
```

This function replaces the initial function with its first derivative until the argument order is reduced to one. Then the third derivative for $3x^3 + x^2$ can be computed with this function.

```
> DD(f, "x", order = 3)
3 * (3 * 2)
```

The gradient of a function can also be computed using the function D().

Definition 2.5 Let $f : \mathbb{R}^n \to \mathbb{R}$ be a differentiable function and $x = (x_1, \ldots, x_n)^\top$ $\in \mathbb{R}^n$. Then the vector

$$\nabla f(x) \overset{\text{def}}{=} \left\{ \frac{\partial f}{\partial x_1}(x), \ldots, \frac{\partial f}{\partial x_n}(x) \right\}^\top$$

is called the *gradient* of f at point x.

If f maps its arguments to a multidimensional space, then we consider the Jacobian matrix as a generalisation of the gradient.

Definition 2.6 Let $F : \mathbb{R}^n \to \mathbb{R}^m$ be a differentiable function with $F = (f_1, \ldots, f_m)$ and the coordinates $x = (x_1, \ldots, x_n)^\top \in \mathbb{R}^n$. Then the *Jacobian matrix* at a point $x \in \mathbb{R}^n$ is defined as follows:

$$J_F(x) \overset{\text{def}}{=} \left\{ \frac{\partial f_i(x)}{\partial x_j} \right\}_{i=1,\ldots,m;\ j=1,\ldots,n}.$$

If one is interested in the second derivatives of a function, the Hessian matrix is important.

Definition 2.7 Let $f : \mathbb{R}^n \to \mathbb{R}$ be twice continuously differentiable. Then the *Hessian matrix* of f at a point x is defined as follows:

$$H_f(x) \overset{\text{def}}{=} \left\{ \frac{\partial^2 f}{\partial x_i \partial x_j}(x) \right\}_{i,j=1,\ldots,n}. \tag{2.19}$$

Now consider the function $f : \mathbb{R}^2 \to \mathbb{R}$ that maps x_1 and x_2 coordinates to the square of their Euclidean norm.

```
> f    = expression(x^2 + y^2)      # function
> grad = c(D(f, "x"), D(f, "y"))    # gradient vector
> grad
[[1]]
2 * x

[[2]]
2 * y
```

If it is necessary to have the gradient as a function that can be evaluated, the function deriv(f, name, function.arg = NULL, hessian = FALSE) should be used. The function argument f is the function (as an object of mode expression) and the argument name identifies the vector with respect to which the derivative will be computed. Furthermore the arguments function.arg specify the parameters of the returned function and hessian indicates whether the second derivatives should be calculated. When `function.arg` is not specified, the return value of `deriv()` is an `expression` and not a function. As an example of the use of the function `deriv()`, consider the above function `f`.

```
> eucld2 = deriv(f,
+    name = c("x","y"),        # variable names +
function.arg = c("x","y"))   # arguments for a function return
> eucld2(2, 2)
[1] 8
attr(,"gradient")
     x y
[1,] 4 4
```

The function `eucld2(x, y)` delivers the value of `f` at `(x, y)` and, as an attribute (see `help(attr)`), the gradient of `f` evaluated at `(x, y)`. If only the evaluated gradient at `(x, y)` should be returned, the function `attr(x, which)` should be used, where `x` is an object and `which` a non-empty character string specifying which attribute is to be accessed.

```
> attr(eucld2(2, 2),"gradient")
     x y
[1,] 4 4
```

If the option `hessian` is set to `TRUE`, the Hessian matrix at a point (x, y) can be retrieved through the call `attr(eucld(2, 2),"hessian")`.

2.3.2 Numerical Differentiation

To develop numerical methods for determining the derivatives of a function at a point x, one uses the Taylor expansion

$$f(x + h) = f(x) + h \cdot f'(x) + \frac{h^2}{2!} f''(x) + \frac{h^3}{3!} f'''(x) + \mathcal{O}(h^3). \qquad (2.20)$$

Only if the fourth derivative of f exists and f is bounded on $[x, x + h]$ the representation in (2.20) is valid. If the Taylor expansion is truncated after the linear term, then (2.20) can be solved for $f'(x)$:

$$f'(x) = \frac{f(x + h) - f(x)}{h} + \mathcal{O}(h). \qquad (2.21)$$

Therefore an approximation for the derivative at point x could be

$$f'(x) \approx \frac{f(x + h) - f(x)}{h}. \qquad (2.22)$$

Another more accurate method uses the Richardson (1911) extrapolation. Redefine the expression in (2.22) with $g(h) = \frac{f(x+h)-f(x)}{h}$. Then (2.21) can be written as

$$f'(x) = g(h) + k_1 h + k_2 h^2 + k_3 h^3 + \cdots , \tag{2.23}$$

where $k_1, k_2, k_3, \ldots$ represent constant terms involving the derivatives at point x. Taylor's theorem holds for all positive h, one therefore can replace h by $h/2$:

$$f'(x) = g\left(\frac{h}{2}\right) + k_1 \frac{h}{2} + k_2 \frac{h^2}{4} + k_3 \frac{h^3}{8} + \cdots . \tag{2.24}$$

Now (2.23) can be subtracted from (2.24) times two. Then the term involving k_1 is eliminated:

$$f'(x) = 2g\left(\frac{h}{2}\right) - g(h) + k_2 \left(\frac{h^2}{2} - h^2\right) + k_3 \left(\frac{h^3}{4} - h^3\right) + \cdots .$$

Therefore $f'(x)$ can be rewritten as follows:

$$f'(x) = 2g\left(\frac{h}{2}\right) - g(h) + \mathcal{O}(h^2).$$

This process can be continued to obtain formulae of higher order. In R, the package numDeriv provides some functions that use these methods to differentiate a function numerically. For example, the function grad() calculates a numerical approximation to the gradient of func at the point x. The argument method can be "simple" or "Richardson". If the method argument is simple, a formula as in (2.22) is applied. Then only the element eps of methods.args is used (equivalent to the above h in (2.22)). The method "Richardson" uses the Richardson extrapolation. Consider the function $f(x_1, x_2, x_3) = \sqrt{x_1^2 + x_2^2 + x_3^2}$, which has the gradient

$$\nabla f(x) = \left(\frac{x_1}{\sqrt{x_1^2 + x_2^2 + x_3^2}}, \frac{x_2}{\sqrt{x_1^2 + x_2^2 + x_3^2}}, \frac{x_3}{\sqrt{x_1^2 + x_2^2 + x_3^2}}\right)^\mathsf{T} .$$

The gradient of f represents the normalised coordinates of a vector with respect to the Euclidean norm. The evaluation of the gradient using grad(), e.g. at the point $(1, 0, 0)$, would be calculated by

```
> require(numDeriv)
> f = function(x){sqrt(sum(x^2))}
> grad(f,
+    x        = c(1,0,0),      # point at which to compute the gradient
+    method ="Richardson") # method to use for the approximation
[1] 1 0 0
> grad(f,
+    x        = c(1,0,0),      # point at which to compute the gradient
+    method ="simple")      # method to use for the approximation
[1] 1e+00 5e-05 5e-05
```

It could also be interesting to compute numerically the Jacobian or the Hessian matrix of a function $F : \mathbb{R}^n \to \mathbb{R}^m$.

In R, the function `jacobian(func,x,...)` can be used to compute the Jacobian matrix of a function `func` at a point `x`. As with the function `grad()`, the function `jacobian()` uses the Richardson extrapolation by default. Consider the following example, where the Jacobian matrix of $f(x) = \{\sin(x_1 + x_2), \cos(x_1 + x_2)\}$ at the point $(0, 2\pi)$ is computed:

```
> require(numDeriv)
> f1 = function(x){c(sin(sum(x)), cos(sum(x)))}
> jacobian(f1, x = c(0, 2 * pi))
      [,1] [,2]
[1,]    1    1
[2,]    0    0
```

The Hessian matrix is symmetric and can be computed in R with `hessian(func,x,...)`. For example, consider $f(x_1, x_2, x_3) = \sqrt{x_1^2 + x_2^2 + x_3^2}$ as above, which maps the coordinates of a vector to their Euclidean norm. The following computation provides the Hessian matrix at the point $(0, 0, 0)$.

```
> f = function(x){sqrt(sum(x^2))}
> hessian(func, c(0, 0, 0))
          [,1]        [,2]        [,3]
[1,] 194419.75 -56944.23 -56944.23
[2,] -56944.23 194419.75 -56944.23
[3,] -56944.23 -56944.23 194419.75
```

From the definition of the Euclidean norm, it would make sense for f to have a minimum at $(0, 0, 0)$. The above information can be used to check whether f has a local minimum at $(0, 0, 0)$. In order to check this, two conditions have to be fulfilled. The gradient at $(0, 0, 0)$ has to be the zero vector and the Hessian matrix should be positive definite (see Canuto and Tabacco 2010 for further information on the calculation of local extreme values using the Hessian matrix). The second condition can be restated by using the fact that a positive definite matrix has only positive eigenvalues. Therefore, the second condition can be checked by computing the eigenvalues of the above Hessian matrix and the first condition can be checked using the `grad()` function.

```
> f = function(x){sqrt(sum(x^2))}
> grad(f, x = c(0, 0, 0))          # gradient at the
[1] 0 0 0                          # optimum point
> hessm = hessian(func, x = c(0, 0, 0))  # Hessian matrix
> eigen(hessm)$values             # eigenvalues
[1] 251364.0 251364.0  80531.3
```

This output shows that the gradient at $(0, 0, 0)$ is the zero vector and the eigenvalues are all positive. Therefore, as expected, the point $(0, 0, 0)$ is a local minimum of f.

2.3.3 Automatic Differentiation

For a function $f : \mathbb{R}^n \to \mathbb{R}^m$, automatic differentiation, which is also called algorithmic differentiation or computational differentiation, is a technique employed to evaluate derivatives based on the chain rule. As the derivatives of the elementary functions, such as exp, log, sin, cos, etc., are already known, the derivative of f can be an automatically assembled from these known elementary partial derivatives by employing the chain rule.

Automatic differentiation is different from two other methods of differentiation, symbolic differentiation and numerical differentiation.

The main difference between automatic differentiation and symbolic differentiation is that the latter focuses on the symbolic expression of formulae and the former concentrates on evaluation. The disadvantages of symbolic differentiation is in both taking up too much memory for the computation, and generating unnecessary expressions associated with the computation. Consider, for example, the symbolic differentiation of

$$f(x) = \prod_{i=1}^{10} x_i = x_1 \cdot x_2 \cdots \cdot x_{10}.$$

The corresponding gradient in symbolic style is

$$\nabla f(x) = \left(\frac{\partial f}{\partial x_1}, \frac{\partial f}{\partial x_2}, \dots, \frac{\partial f}{\partial x_{10}} \right)$$

$$= (x_2 \cdot x_3 \cdots \cdot x_{10}, x_1 \cdot x_3 \cdots \cdot x_{10}, \dots, x_1 \cdot x_2 \cdots \cdot x_9).$$

If the number of variables becomes large, then the expression will use a tremendous amount of memory and have a very tedious representation.

In automatic differentiation, all arguments of the function are redefined as dual numbers, $x_i + x_i' \varepsilon$, where ε has the property that $\varepsilon^2 \approx 0$. The change in x_i is $x_i' \varepsilon$, for all i. Therefore, automatic differentiation for this function looks like

$$\nabla f(x) = \prod_{i=1}^{10} x_i + \varepsilon \left(x_1' \prod_{i=2}^{10} x_i + \cdots + x_j' \prod_{i \neq j}^{10} x_i + \cdots + x_{10}' \prod_{i=1}^{9} x_i \right).$$

Automatic differentiation is more accurate than numerical differentiation. Numerical differentiation (or divided differences) uses

$$f'(x) \approx \frac{f(x+h) - f(x)}{h},$$

or

$$f'(x) \approx \frac{f(x+h) - f(x-h)}{2h}.$$

It is obvious that the accuracy of this type of differentiation is related to the choice
of h. If h is small, then the method of divided differences has errors introduced by
rounding off the floating point numbers. If h is large, then the formula disobeys
the essence of this method, which assumes that h tends to zero. Also, the method
of divided differences introduces truncation errors by neglecting the terms of order
$\mathcal{O}(h^2)$, something which does not happen in automatic differentiation.

Automatic differentiation has two operation modes: forward and reverse. For
forward mode, the algorithm starts by evaluating the derivatives of every elementary
function, the function arguments itself, of f at the given points. In each intermediate
step, the derivatives are combined to reproduce the derivatives of more complicated
functions. The last step merely assembles the evaluations from the results of the
computations already performed, employing the chain rule. For example, we use the
forward mode to evaluate the derivative of $f(x) = (x + x^2)^3$: the pseudocode can
be summarised as

```
function(y, y')=f'(x, x')
  s1  = x * x;
  s1' = 2 * x * x';
  s2  = x + s1;
  s2' = x' + s1';
  y  = s2 * s2 * s2;
  y' = 3 * s2 * s2 * s2'
end
```

where f' represents the derivative, i.e. $\partial f/\partial x$. Therefore, let us evaluate the derivative
of $f(x) = (x + x^2)^3$ at the point $x = 2$ with the forward mode.

$$s_1 = x \cdot x = 2 \cdot 2 = 4,$$
$$s_1' = 2 \cdot x \cdot x' = 2 \cdot 2 \cdot 1 = 4,$$
$$s_2 = x + s_1 = 2 + 4 = 6,$$
$$s_2' = x' + s_1' = 1 + 4 = 5,$$
$$y = s_2 \cdot s_2 \cdot s_2 = 6 \cdot 6 \cdot 6 = 216,$$
$$y' = 3 \cdot s_2 \cdot s_2 \cdot s_2' = 3 \cdot 6 \cdot 6 \cdot 5 = 540.$$

For reverse mode, the programme performs the computation in the reverse direction.
We need to set $\bar{v} = dy/dv$, then $\bar{y} = dy/dy = 1$. For the same example as before,
where the derivative at $x = 2$ is evaluated, it looks as

$$\bar{s}_2 = 3 \cdot s_2^2 = 3 \cdot 36 = 108,$$
$$\bar{s}_1 = 3 \cdot s_2^2 = 3 \cdot 36 = 108,$$
$$\bar{x} = \bar{s}_2 + 4\bar{s}_1 = 108 + 4 \cdot 108 = 540.$$

Two examples are implemented in R using the package radx developed by Annamalai (2010). This package is not available on CRAN, therefore is installed via

function `install_github` (provided by package `devtools`) from GitHub from repository `radx` by `quantumelixir`.

Example 2.5 Evaluate the first-order derivative of

$$f(x) = \left(x + x^2\right)^3, \quad for \ x = 2. \tag{2.25}$$

```
> require(devtools)
> # install_github("radx","quantumelixir")   # installs from GitHub
> require(radx)                                # not provided by CRAN
> f = function(x) {(x^2 + x)^3}                # function
> radxeval(f,                                  # automatic differ.
+    point = 2,                                 # point at which to eval.
+    d     = 1)                                 # order of differ.
       [,1]
[1,]   540
```

The upper computation illustrates that the value of the first derivative of the function (2.25) at $x = 2$ is equal to 540.

Example 2.6 Evaluate the first and second derivatives of the vector function

$$f_1(x, y) = 1 - 3y + \sin(3\pi y) - x,$$
$$f_2(x, y) = y - \sin(3\pi x)/2,$$

at $(x = 3, y = 5)$.

```
> f = function(x, y){                          # multidimensional function
+    c(1 - 3 * y + sin(3 * pi * y) - x,
+      y - sin(3 * pi * x) / 2)
+ }
> radxeval(f,
+    point = c(3, 5),                           # point at which to evaluate
+    d     = 1)                                 # 1st order of differentiation
        [,1]      [,2]
[1,]  -1.00000  4.712389
[2,] -12.42478  1.000000
> radxeval(f,
+    point = c(3, 5),                           # point at which to evaluate
+    d     = 2)                                 # 2nd order of differentiation
             [,1]          [,2]
[1,]  0.00000e+00  4.894984e-14
[2,]  0.00000e+00  0.000000e+00
[3,] -4.78741e-13  0.000000e+00
```

2.4 Root Finding

A root is the solution to a system of equations or an optimisation problem. In both cases, one tries to find values for the arguments of the function such that the value of the function is zero. In the case of an optimisation problem, this is done for the first derivative of the objective function.

2.4.1 Solving Systems of Linear Equations

Let K denote either the set of real numbers, or the set of complex numbers. Suppose $a_{ij}, b_i \in K$ with $i = 1, \ldots, n$ and $j = 1, \ldots, p$. Then the following system of equations is called a system of linear equations:

$$\begin{cases} a_{11}x_1 + \ldots + a_{1p}x_p = b_1 \\ \vdots \qquad\qquad\qquad \vdots \\ a_{n1}x_1 + \ldots + a_{np}x_p = b_n \end{cases}$$

For a matrix $\mathcal{A} = (a_{ij}) \in K^{n \times p}$ and two vectors $x = (x_1, \ldots, x_p)^\top$ and $b = (b_1, \ldots, b_n)^\top$, the system of linear equations can be rewritten in matrix form:

$$\mathcal{A}x = b. \tag{2.26}$$

Let $\mathcal{A}^e_{n \times (p+1)}$ be the extended matrix, i.e. the matrix whose last column is the vector of constants b, and otherwise is the same as $\mathcal{A}$. Then (2.26) can be solved if and only if the rank of $\mathcal{A}$ is the same as the rank of $\mathcal{A}^e$. In this case b can be represented by a linear combination of the columns of $\mathcal{A}$. If (2.26) can be solved and the rank of $\mathcal{A}$ equals $n = p$, then there exists a unique solution. Otherwise (2.26) might have no solution or infinitely many solutions, see Greub (1975).

The Gaussian algorithm, which transforms the system of equations by elementary transformations to upper triangular form, is frequently applied. The solution can be computed by back-substitution. The Gaussian algorithm decomposes $\mathcal{A}$ into the matrices $\mathcal{L}$ and $\mathcal{U}$, the so-called $\mathcal{LU}$ decomposition (see Braun and Murdoch (2007) for further details). $\mathcal{L}$ is a lower triangular matrix and $\mathcal{U}$ is an upper triangular matrix with the following form:

$$\mathcal{L} = \begin{pmatrix} 1 & 0 & \ldots & 0 \\ l_{21} & 1 & \ldots & \vdots \\ \vdots & & \ddots & 0 \\ l_{n1} & l_{n2} & \ldots & 1 \end{pmatrix}, \qquad \mathcal{U} = \begin{pmatrix} u_{11} & u_{12} & \ldots & u_{1n} \\ 0 & u_{22} & \ldots & u_{2n} \\ \vdots & & \ddots & \vdots \\ 0 & \ldots & 0 & u_{nn} \end{pmatrix}.$$

Then (2.26) can be rewritten as

$$Ax = \mathcal{LU}x = b. \tag{2.27}$$

Now the system in (2.26) can be solved in two steps. First define $\mathcal{U}x = y$ and solve $\mathcal{L}y = b$ for y by forward substitution. Then solve $\mathcal{U}x = y$ for x by back-substitution. In R, the function $\texttt{solve(A,b)}$ uses the $\mathcal{LU}$ decomposition to solve a system of linear equations with the matrix A and the right side b. Another method that can be used in R to solve a system of linear equations is the $\mathcal{QR}$ decomposition, where the matrix $\mathcal{A}$ is decomposed into the product of an orthogonal matrix $\mathcal{Q}$ and an upper triangular matrix $\mathcal{R}$. One uses the function $\texttt{qr.solve()}$ to compute the solution of a system of linear equations using the $\mathcal{QR}$ decomposition. In contrast to the $\mathcal{LU}$ decomposition, this method can be applied even if $\mathcal{A}$ is not a square matrix. The next example shows how to solve a system of linear equations in R using $\texttt{solve()}$.

Example 2.7 Solve the following system of linear equations in R with the Gaussian algorithm and back-substitution,

$$Ax = b,$$

$$\mathcal{A} = \begin{pmatrix} 2 & -\frac{1}{2} & -\frac{1}{2} & 0 \\ -\frac{1}{2} & 0 & 2 & -\frac{1}{2} \\ -\frac{1}{2} & 2 & 0 & -\frac{1}{2} \\ 0 & -\frac{1}{2} & -\frac{1}{2} & 2 \end{pmatrix},$$

$$b = (0, 3, 3, 0)^{\top},$$

$$\mathcal{A}^e = \begin{pmatrix} 2 & -\frac{1}{2} & -\frac{1}{2} & 0 & 0 \\ -\frac{1}{2} & 0 & 2 & -\frac{1}{2} & 3 \\ -\frac{1}{2} & 2 & 0 & -\frac{1}{2} & 3 \\ 0 & -\frac{1}{2} & -\frac{1}{2} & 2 & 0 \end{pmatrix}.$$

The upper system of linear equations is solved first by hand and then the example is computed in R for verification. This system of linear equations is not difficult to solve with the Gaussian algorithm. First, one finds the upper triangular matrix

$$\mathcal{U}^e = \begin{pmatrix} 2 & -\frac{1}{2} & -\frac{1}{2} & 0 & 0 \\ 0 & \frac{15}{8} & -\frac{1}{8} & -\frac{1}{2} & 3 \\ 0 & 0 & \frac{28}{15} & -\frac{8}{15} & \frac{16}{5} \\ 0 & 0 & 0 & \frac{12}{7} & \frac{12}{7} \end{pmatrix}.$$

Second, one uses back-substitution to obtain the final result, that $(x_1, x_2, x_3, x_4)^{\top} = (1, 2, 2, 1)^{\top}$. Then the solution of this system of linear equations in R is presented. Two parameters are required: the coefficient matrix A and the vector of constraints b.

```
> A = matrix(                    # coefficient matrix
+    c(    2, -1/2, -1/2,      0,
+       -1/2,    0,    2, -1/2,
+       -1/2,    2,    0, -1/2,
```

```
+            0, -1/2, -1/2,     2),
+      ncol = 4, nrow = 4, byrow = TRUE)
> b = c(0, 3, 3, 0)                    # vector of constants
> solve(A, b)
[1] 1 2 2 1                            # x1, x2, x3, x4
```

The manually found solution for the system coincides with the solution found in R.

2.4.2 Solving Systems of Nonlinear Equations

A system of nonlinear equations is represented by a function $F = (f_1, \ldots, f_n) :$ $\mathbb{R}^n \to \mathbb{R}^n$. Any nonlinear system has a general extended form

$$
\begin{cases}
f_1(x_1, \ldots, x_n) = 0, \\
\vdots \qquad\qquad \vdots \\
f_n(x_1, \ldots, x_n) = 0.
\end{cases}
$$

There are many different numerical methods for solving systems of nonlinear equations. In general, one distinguishes between gradient and non-gradient methods. In the following, the *Newton method*, or the *Newton–Raphson method*, is presented. To get a better illustration of the idea behind the Newton method, consider a continuous differentiable function $F : \mathbb{R} \to \mathbb{R}$, where one tries to find x^* with $F(x^*) = 0$ and $\left.\frac{\partial F(x)}{\partial x}\right|_{x=x^*} \neq 0$. Start by choosing a starting value $x_0 \in \mathbb{R}$ and define the tangent line

$$
p(x) = F(x_0) + \left.\frac{\partial F(x)}{\partial x}\right|_{x=x_0} (x - x_0). \tag{2.28}
$$

Then the tangent line $p(x)$ is a good approximation to F in a sufficiently small neighbourhood of x_0. If $\left.\frac{\partial F(x)}{\partial x}\right|_{x=x_0} \neq 0$, the root x_1 of p in (2.28) can be computed as follows:

$$
x_1 = x_0 - \frac{F(x_0)}{\left.\frac{\partial F(x)}{\partial x}\right|_{x=x_0}}.
$$

With the new value x_1, the rule can be applied again. This procedure can be applied iteratively and under certain theoretical conditions the solution should converge to the actual root. Figure 2.2 demonstrates the Newton method for $f(x) = x^2 - 4$ with the starting value $x_0 = 6$.

The Fig. 2.2 was computed using the function newton.method(f, init, ...) from the package animation, where f is the function of interest and init is the starting value for the iteration process. The function provides an illustration of the iterations in Newton's method (see help(newton.method) for further details). The function uniroot() searches in an interval for a root of a function and returns

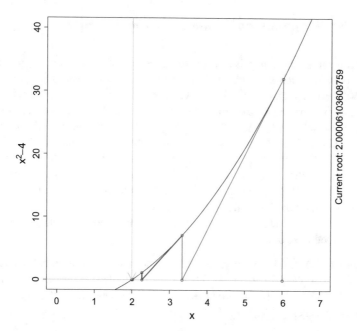

Fig. 2.2 Illustration of the iteration steps of Newton's method to find the root of $f(x) = x^2 - 4$ with $x_0 = 6$. Q BCS_Newton

only one root, even if several roots exist within the interval. At the boundaries of the interval, the sign of the value of the function must change.

```
> f = function(x){              # objective function
+     -x^4 - cos(x) + 9 * x^2 - x - 5
+ }
> uniroot(f,
+     interval = c(0, 2))$root    # root in [0, 2]
[1] 0.8913574
> uniroot(f,
+     interval = c(-3, 2))$root   # root in [-3, 2]
[1] -2.980569
> uniroot(f,
+     interval = c(0, 3))$root    # f(0) and f(3) negative
Error in uniroot(f, c(0, 3)) :
  Values of f() at the boundaries have same sign
```

For a real or complex polynomial of the form $p(x) = z_1 + z_2 \cdot x + \ldots + z_n \cdot x^{n-1}$, the function polyroot(z), with z being the vector of coefficients in the increasing order, computes a root. The algorithm does not guarantee it will find all the roots of the polynomial.

```
> z = c(0.2567, 0.1570, 0.0821, -0.3357, 1) # coefficients
> round(polyroot(z), digits = 2)             # complex roots
[1]  0.59+0.60i -0.42+0.44i -0.42-0.44i  0.59-0.60i
```

2.4.3 Maximisation and Minimisation of Functions

The maximisation and minimisation of functions, or optimisation problems, contain
two components, such as an objective function $f(x)$ and constraints $g(x)$. Optimisation problems can be classified into two categories, according to the existence of
constraints. If there are constraints affiliated with the objective function, then it is
a constrained optimisation problem; otherwise, it is a unconstrained optimisation
problem. This section introduces six different optimisation techniques. The first four
are the golden ratio search method, the Nelder–Mead method, the BFGS method, and
the conjugate gradient method for unconstrained optimisation. The two other optimisation techniques, linear programming (LP) and nonlinear programming (NLP),
are used for constrained optimisation problems.

First, one needs to define the concepts of local and global extrema, which will be
frequently used later on.

Definition 2.8 A real function f defined on a domain M has a global maximum at
x_{opt} if $f(x_{opt}) \geq f(x)$ for all x in M. Then $f(x_{opt})$ is called the maximum value of the
function. Analogously, the function has a global minimum at x_{opt} if $f(x_{opt}) \leq f(x)$
for all x in M. Then $f(x_{opt})$ is called the minimum value of the function.

Definition 2.9 If the domain M is a metric space, then f is said to have a local
maximum at x_{opt} if there exists some $\epsilon > 0$ such that $f(x_{opt}) \geq f(x)$ for all x in M
within a distance of ϵ from x_{opt}. Analogously, the function has a local minimum at
x_{opt} if $f(x_{opt}) \leq f(x)$ for all x in M within ϵ of x_{opt}.

Maxima and minima are not always unique. Consider the function $\sin(x)$, which
has global maxima $f(x_{max}) = 1$ and global minima $f(x_{min}) = -1$ for every $x_{max} =
(0.5 + 2k)\pi$ and $x_{min} = (-0.5 + 2k)\pi$ for $k \in \mathbb{Z}$.

Example 2.8 The following function possesses several local maxima, local minima,
global maxima and global minima.

$$f(x, y) = 0.03 \sin(x) \sin(y) - 0.05 \sin(2x) \sin(y)$$
$$+ 0.01 \sin(x) \sin(2y) + 0.09 \sin(2x) \sin(2y). \tag{2.29}$$

The function is plotted in Fig. 2.3 with highlighted extrema.

Golden ratio search method

The golden ratio section search method was proposed in Kiefer (1953).
This method is frequently employed for solving optimisation problems with one-
dimensional uni-modal objective functions, and belongs to the group of non-gradient
methods. A very common algorithm for this method is the following one:

1. Define upper bound x_U, lower bound x_L, $x_U > x_L$ and ε.
2. Set $r = (\sqrt{5} + 1)/2$, $d = (x_U - x_L)r$.
3. Choose x_1 and $x_2 \in [x_L, x_U]$. Set $x_1 = x_L + d$ and $x_2 = x_U - d$.

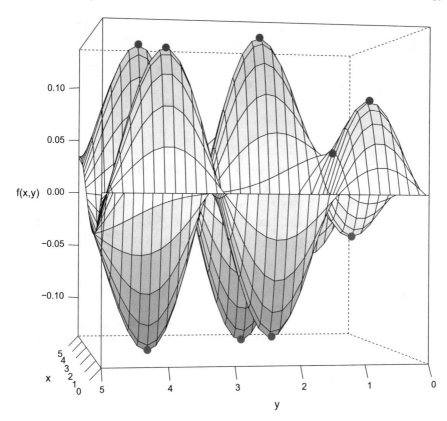

Fig. 2.3 3D plot of the function (2.29) with maxima and minima depicted by points.
Q BCS_Multimodal

4. Stop if $|f(x_U) - f(x_L)| < \varepsilon$ or $|x_U - x_L| < \varepsilon$. If $f(x_1) < f(x_2) \Rightarrow x_{min} = x_1$, otherwise $x_{min} = x_2$.
5. Update x_U or x_L if $|f(x_U) - f(x_L)| \geq \varepsilon$ or $|x_U - x_L| \geq \varepsilon$. If $f(x_1) > f(x_2) \Rightarrow x_U = x_2$, otherwise $x_L = x_1$.
6. Return to step 2.

The algorithm first defines an initial search interval d. The length of the interval depends on the difference between the upper and lower bounds $x_U - x_L$ and the 'Golden Ratio' $r = \frac{1+\sqrt{5}}{2}$. The points x_1 and x_2 will decrease the length of the search interval. If the tolerance level ε for the search criteria ($|f(x_U) - f(x_L)|$ or $|x_U - x_L|$) is satisfied, the process stops. But if the search criteria is still greater than or equal to the tolerance level, the bounds are updated and the algorithm starts again.

Example 2.9 Apply the golden ratio search method to find the maximum of

$$f(x) = -(x - 3)^2 + 10. \tag{2.30}$$

```
> require(stats)
> f = function(x){-(x - 3)^2 + 10}    # function
> optimize(f,                         # objective function
+    interval = c(-10, 10),           # interval
+    tol        = 0.0001,             # level of the tolerance
+    maximum = TRUE)                  # to find maximum
$maximum
[1] 3

$objective
[1] 10
```

The argument `tol` defines the convergence criterion for the results. The function reaches its global maximum at $x_{opt} = 3$, which is easily derived by solving the first-order condition $-2x_{opt} + 6 = 0$ for x_{opt} and computing the value $f(x_{opt})$. For a maximum at x_{opt}, one should have $\frac{\partial^2 f(x)}{\partial x^2} < 0$ and $\left.\frac{\partial^2 f(x)}{\partial x^2}\right|_{x=x_{opt}} = -2$. Therefore $x_{opt} = 3$, which is verified in R with the code from above.

Nelder–Mead method

This method was proposed in Nelder and Mead (1965) and is applied frequently in multivariate unconstrained optimisation problems. It is a direct method, where the computation does not use gradients. The main idea of the Nelder–Mead method is briefly explained below and a graph for a two-dimensional input case is shown in Fig. 2.4.

1. Choose x_1, x_2, x_3 such that $f(x_1) < f(x_2) < f(x_3)$ and set ϵ_x and/or ϵ_f.
2. Stop if $\|x_i - x_j\| < \epsilon_x$ and/or $\|f(x_i) - f(x_j)\| < \epsilon_f$, for $i \neq j$, i, $j \in \{1, 2, 3\}$ and set $x_{min} = x_1$.
3. Else, compute $z = \frac{1}{2}(x_1 + x_2)$ and $d = 2z - x_3$.

   If $f(x_1) < f(d) < f(x_2) \Rightarrow x_3 = d$.
   If $f(d) \leq f(x_1)$, compute $k = 2d - z$.
      If $f(k) < f(x_1) \Rightarrow x_3 = k$.
      Else, $x_3 = d$.
   If $f(x_3) > f(d) \geq f(x_2) \Rightarrow x_3 = d$.
   Else, compute $t = [t|f(t) = \min\{f(t_1), f(t_2)\}]$, where $t_1 = \frac{1}{2}(x_3 + z)$ and $t_2 = \frac{1}{2}(d + z)$.
      If $f(t) < f(x_3) \Rightarrow x_3 = t$.
      Else, $x_3 = s = 1/2(x_1 + x_3)$ and $x_2 = z$.

4. Return to step 2.

In general, the Nelder–Mead algorithm works with more than three initial guesses. The starting values x_i are allowed to be vectors. In the iteration procedure one tries to improve the initial guesses step by step. The worst guess x_3 will be replaced by better values until the convergence criterion for the values ϵ_f of the function or the arguments ϵ_x of the function is met. Next we will give an example of how to use the Nelder–Mead method to find extrema of a function in R (Fig. 2.5).

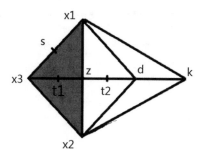

Fig. 2.4 Algorithm graph for the Nelder–Mead method. The variables x_1, x_2 and x_3 are the search region at the specific iteration step. All other variables, d, k, s, t_1 and t_2, are possible updates for one x_i

Example 2.10 The function to be minimized is the Rosenbrock function, which has an analytic solution with global minimum at $(1, 1)$ and a global minimum value $f(1, 1) = 0$.

$$f(x_1, x_2) = 100(x_2 - x_1^2)^2 + (1 - x_1)^2. \tag{2.31}$$

```
> require(neldermead)
> f = function(x){
+     100 * (x[2] - x[1]^2)^2 + (1 - x[1])^2 # Rosenbrock function
+ }
> fNM = fminsearch(fun = f,
+     x0      = c(-1.2, 1),                  # starting point
+     verbose = FALSE)
> neldermead.get(fNM, key = "xopt")         # optimal x-values
         [,1]
[1,] 1.000022
[2,] 1.000042
> neldermead.get(fNM, key = "fopt")         # optimal function value
[1] 8.177661e-10
```

The upper computation illustrates that the numerical solution by the Nelder–Mead method is close to the analytical solution for the Rosenbrock function (2.31). The errors of the numerical solution are negligibly small.

BFGS method

This frequently used method for multivariate optimisation problems was proposed independently in Broyden (1970), Fletcher (1970), Goldfarb (1970) and Shanno (1970). BFGS stands for the first letters of each author, in alphabetical order. The main idea of this method originated from Newton's method, where the second-order Taylor expansion for a twice differentiable function $f : \mathbb{R}^n \to \mathbb{R}$ at $x = x_i \in \mathbb{R}^n$ is employed, such that

$$f(x) = f(x_i) + \nabla f^\top(x_i)q + \frac{1}{2}q^\top H(x_i)q,$$

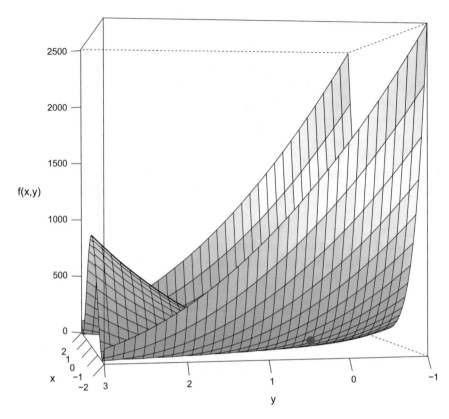

Fig. 2.5 Plot for the Rosenbrock function with its minimum depicted by a point.
⌕ BCS_Rosenbrock

where $q = x - x_i$, and $\nabla f(x_i)$ is the value of the partial derivative of f at the point x_i, and $H(x_i)$ is the Hessian matrix. Employing the first-order condition, one obtains

$$\nabla f(x) = \nabla f(x_i) + H(x_i)q = 0,$$

hence, if $H(x_i)$ is invertible, then

$$q = x - x_i = -H^{-1}(x_i)\nabla f(x_i),$$
$$x_{i+1} = x_i - H^{-1}(x_i)\nabla f(x_i).$$

The recursion will converge quadratically to the optimum. The problem is that Newton's method requires the computation of the exact Hessian at each iteration, which is computationally expensive. Therefore, the BFGS method overcomes this disadvantage with an approximation of the Hessian's inverse obtained from the following optimisation problem,

$$H(x_i) = \arg\min_H \|H^{-1} - H^{-1}(x_{i-1})\|_W, \tag{2.32}$$

$$\text{subject to:} \qquad \mathcal{B}^{-1} = (H^{-1})^\top,$$
$$\mathcal{B}^{-1}(\nabla f_i - \nabla f_{i-1}) = x_i - x_{i-1}.$$

The weighted Frobenius norm, denoted by $\|\cdot\|_W$, and the matrix W are, respectively,

$$\|H^{-1} - H^{-1}(x_{i-1})\|_W = \|W^{\frac{1}{2}}\{H^{-1} - H^{-1}(x_{i-1})\}W^{\frac{1}{2}}\|,$$
$$W(\nabla f_i - \nabla f_{i-1}) = x_i - x_{i-1}.$$

Equation (2.32) has a unique solution such that

$$H^{-1}(x_i) = \mathcal{M}_1 H(x_{i-1})\mathcal{M}_2 + (x_i - x_{i-1})\gamma_{i-1}(x_i - x_{i-1})^\top,$$
$$\mathcal{M}_1 = \mathcal{I} - \gamma_{i-1}(x_i - x_{i-1})(\nabla f_i - \nabla f_{i-1})^\top,$$
$$\mathcal{M}_2 = \mathcal{I} - \gamma_{i-1}(\nabla f_i - \nabla f_{i-1})(x_i - x_{i-1})^\top,$$
$$\gamma_i = \{(\nabla f_i - \nabla f_{i-1})^\top(x_i - x_{i-1})\}^{-1}.$$

Example 2.11 Here, the BFGS method is used to minimise the Rosenbrock function
(2.31) using `optimx` package (see Nash and Varadhan 2011).

```
> require(optimx)
> f       = function(x){100 * (x[2] - x[1]^2)^2 + (1 - x[1])^2}
> fBFGS = optimx(fn = f,                    # objective function
+    par      = c(-1.2, 1),                  # starting point
+    method = "BFGS")                        # optimisation method
> print(data.frame(fBFGS$p1, fBFGS$p2, fBFGS$value))
   fBFGS.p1   fBFGS.p2   fBFGS.value
1 0.9998044 0.9996084 3.827383e-08        # minimum
```

The BFGS method computes the minimum value of the function (2.31) to be $3.83e - 08$ at the minimum point $(0.99, 0.99)$. The outputs `fevals = 127`, `gevals = 38` show the calls of the objective function and the calls of the gradients, respectively. These outputs are close to the exact solution $x_{opt} = (1, 1)$ and $f(x_{opt}) = 0$.

Conjugate gradient method

The conjugate gradient method was proposed in Hestenes and Stiefel (1952) and is widely used for solving symmetric positive definite linear systems. A multivariate unconstrained optimisation problem, like

$$\mathcal{A}x = b, \quad \mathcal{A} \in \mathbb{R}^{n \times n}, \quad \mathcal{A} = \mathcal{A}^\top, \text{ and } \mathcal{A} \text{ positive definite,}$$

can be solved with the Conjugate Gradient Method. The main idea behind this method is to use iterations to approach the optimum of the linear system.

1. Set x_0 and ε, then compute $p_0 = r_0 = b - \mathcal{A}x_0$.
2. Stop if $r_i < \varepsilon$ and set $x_{opt} = x_{i+1}$.
3. Else, compute:

$$\alpha_i = \frac{r_i^\top r_i}{p_i^\top \mathcal{A}p_i},$$

$$x_{i+1} = x_i + \alpha_i p_i,$$

$$r_{i+1} = r_i - \alpha_i \mathcal{A}p_i,$$

$$\beta_i = \frac{r_{i+1}^\top r_{i+1}}{r_i^\top r_i},$$

$$p_{i+1} = r_{i+1} + \beta_i p_i.$$

4. Update x_i, r_i and p_i. Increment i.
5. Return to step 2.

At first, the initial guess x_0 determines the residual r_0 and the initially used basis vector p_0. The algorithm tries to reduce the residual r_i at each step to get to the optimal solution. At the optimum, $0 = b - Ax_i$. The tolerance level t and the final residual should be close to zero. The parameters α_i and β_i are improvement parameters for the next iteration. The parameter α_i directly determines the size of the improvement for the residual and indirectly influences the conjugate vector p_{i+1} used in the next iteration. For β_i, the opposite is true. The final result depends on both parameters.

Example 2.12 To illustrate the Conjugate Gradient method, let us again consider the Rosenbrock function (2.31).

```
> require(optimx)
> f    = function(x){100 * (x[2] - x[1]^2)^2 + (1 - x[1])^2}
> fCG = optimx(fn = f,                         # objective function
+    par     = c(1.2, 1),                       # initial guess (x_0)
+    control = list(reltol = 10^-7),            # relative tolerance
+    method  = "CG")                            # method of optimisation
>
> print(data.frame(fCG$p1, fCG$p2, fCG$value))  # minimum
   fCG.p1    fCG.p2    fCG.value
1  1.030077  1.061209  0.0009036108
```

For the Rosenbrock function, the Conjugate Gradient method delivers the biggest errors, compared to the Nelder–Mead and BFGS methods. All numerical methods which are applied to optimize a function will only approximately find the true solution. The examples above show how the choice of method might influence the accuracy of the result. Worth mentioning is, that in the latter case we changed the initial guess, as the function failed with the same starting value as we took for BFGS method.

Constrained optimisation

Constrained optimisation problems can be categorised into two classes in terms of to the linearity of the objective function and the constraints. A linear programming

(LP) problem has a linear objective function and linear constraints, otherwise it is a nonlinear programming problem (NLP).

LP is a method to find the solution to an optimisation problem with a linear objective function, under constraints in the form of linear equalities and linear inequalities. It has a feasible region defined by a convex polyhedron, which is a set made by the intersection of finitely many half-spaces. These represent linear inequalities. The objective of linear programming is to find a point in the polyhedron where the objective function reaches a minimum or maximum value. A representative LP can be expressed as follows:

$$\arg \max_{x} a^\top x,$$
$$\text{subject to: } Cx \leq b,$$
$$x \geq 0,$$

where $x \in \mathbb{R}^n$ is a vector of variables to be identified, a and b are vectors of known coefficients and C is a known matrix of the coefficients in the constraints. The expression $a^\top x$ is the objective function. The inequalities $Cx \leq b$ and $x \geq 0$ are the constraints, under which the objective function will be optimized.

NLP has an analogous definition as that of the LP problem. The differences between NLP and LP are that the objective function or the constraints in an NLP can be nonlinear functions. The following example is an LP problem (Fig. 2.6).

Example 2.13 Solve the following linear programming optimisation problem with R.

$$\arg \max_{x_1, x_2} 2x_1 + 4x_2, \qquad (2.33)$$
$$\text{subject to: } 3x_1 + 4x_2 \leq 60,$$
$$x_1 \geq 0,$$
$$x_2 \geq 0.$$

For the example in (2.33), the function from the package `Rglpk` (see Theussl 2013) is used to compute the solution in R.

```
> require(Rglpk)
> Rglpk_solve_LP(obj = c(2, 4),         # objective function
+    mat = matrix(c(3, 4), nrow = 1),   # constrains coefficients
+    dir ="<=",                          # type of constrains
+    rhs = 60,                           # constrains vector
+    max = TRUE)                         # to maximise
$optimum                                 # maximum
[1] 60

$solution                                # point of maximum
[1]  0 15

$status                                  # no errors
[1] 0
```

The maximum value of the function (2.33) is 60 and occurs at the point (0, 15).

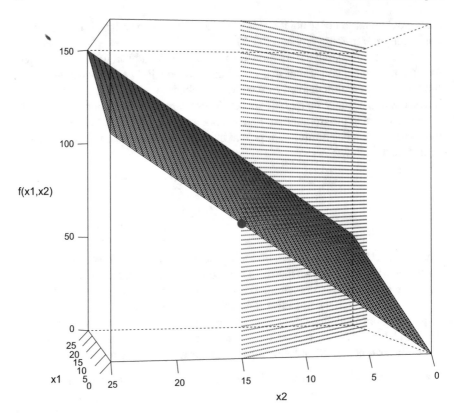

Fig. 2.6 Plot for the linear programming problem with the constraint hyperplane depicted by the grid and the optimum by a point. **Q** BCS_LP

Example 2.14 Next, consider a constrained nonlinear optimisation problem, which can be solved in R using `constrOptim()`. Solve the following nonlinear optimisation problem with R (Fig. 2.7).

$$\arg \max_{x_1, \, x_2} \sqrt{5x_1} + \sqrt{3x_2}, \qquad (2.34)$$

$$\text{subject to: } 3x_1 + 5x_2 \leq 10,$$
$$x_1 \geq 0,$$
$$x_2 \geq 0.$$

```
> require(stats)
> f       = function(x){
+    sqrt(5 * x[1]) + sqrt(3 * x[2])      # objective function
+ }
> A       = matrix(c(-3, -5), nrow = 1,
+    ncol = 2, byrow = TRUE)              # coefficients matrix
> b       = c(-10)                        # vector of constraints
```

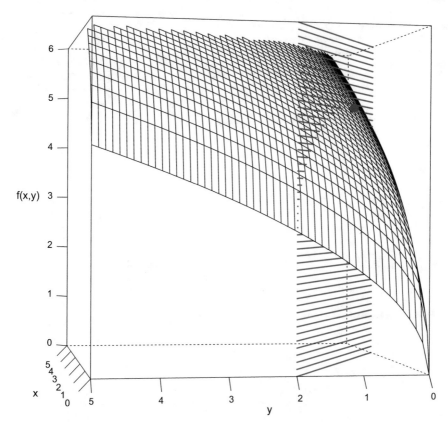

Fig. 2.7 Plot for the objective function with its constraint from (2.34) and the optimum depicted by the point. ⚙ BCS_NLP

```
> answer = constrOptim(f = f,            # objective function
+    theta   = c(1, 1),                   # initial guess
+    grad    = NULL,                      # no gradient provided
+    ui      = A,
+    ci      = b,                         # vector of constrains
+    control = list(fnscale = -1))        # to maximise
> c(answer$par, answer$value)            # optimum
[1] 2.4510595 0.5293643 4.7609523
```

The upper computation illustrates that the maximum value of the function (2.34) is 4.7610, and occurs at the point $(2.4511, 0.5294)$. `answer$function` equal to 170 means that the objective function has been called 170 times.

Chapter 3
Combinatorics and Discrete Distributions

The roll of the dice will never abolish chance.

— Stéphane Mallarmé

In the second half of the nineteenth century, the German mathematician Georg Cantor developed the greater part of today's set theory. At the turn of the nineteenth and twentieth centuries, Ernst Zermelo, Bertrand Russell, Cesare Burali-Forti and others found contradictions in the nonrestrictive set formation: For every property there is a unique set of individuals, which have that property, see Johnson (1972). This so called 'naïve comprehension principle' produced inconsistencies, illustrated by the famous Russell paradox, and was therefore untenable. Ernst Zermelo in 1908 gave an axiomatic system which precisely described the existence of certain sets and the formation of sets from other sets. This Zermelo–Fraenkel set theory is still the most common axiomatic system for set theory. There are 9 axioms, amongst others, that deal with set equality, regularity, pairing sets, infinity, and power sets. Since these axioms are very theoretical, we refer the interested reader to Jech (2003). Later, further axioms were added in order to be able to universally interpret all mathematical objects or constructs, making set theory a fundamental discipline of mathematics. It also plays a major role for computational statistics since it mostly uses basic functions, which constitute set theoretical relations.

3.1 Set Theory

3.1.1 Creating Sets

In mathematics, the most famous sets are

© Springer International Publishing AG 2017
W.K. Härdle et al., *Basic Elements of Computational Statistics*,
Statistics and Computing, DOI 10.1007/978-3-319-55336-8_3

- $\mathbb{N}$: the set of natural numbers, i.e. $\{1, 2, 3, 4 \ldots\}$.
- $\mathbb{Z}$: the set of integer numbers, i.e. $\{\cdots - 3, -2, -1, 0, 1, 2, 3 \ldots\}$.
- $\mathbb{Q}$: the set of rational numbers. An example of a finite set of rational numbers is, e.g., $C = \{-3.5, 0.01, \frac{19}{20}\} \subset \mathbb{Q}$.
- $\mathbb{R}$: the set of real numbers. An example of a finite set of real numbers is, e.g. $D = \{1, 2.5, -3, \sqrt{2}\} \subset \mathbb{R}$.
- $\mathbb{C}$: the set of complex numbers. An example of a finite set of complex numbers is, e.g. $E = \{2 + 3i, 1 - i\} \in \mathbb{C}$, with $i^2 = -1$.

For each set there is a cardinal number which stands for the magnitude or number of elements in the set and allows describing infinite sets. For the set of natural numbers, for example, its cardinal number is $\aleph_0$ ('Aleph null'). In R, of course, one cannot create infinite sets due to its limited storage capacity.

The sets above are named by a fixed character or letter, whereas other sets in the literature are labelled arbitrarily with a Latin or Greek letter, for example, a, b, N, M, Γ.

Using R, a set can be defined by enumerating its elements or by stating its form, as in the following.

```
> A = c(1, 2, 3, 4); A
[1] 1 2 3 4
> B = seq(from = -3, to = 3, by = 1); B
[1] -3 -2 -1 0 1 2 3
```

Most of the basic R objects containing several elements, such as an array, a matrix, or a data frame, are sets.

3.1.2 Basics of Set Theory

After the creation of a set, the next step is to manipulate the set in useful ways. One possible goal could be selecting a specific subset. A subset of a set M is another set M_1 whose elements a are also elements of M, i.e. $a \in M_1$ implies $a \in M$. There are several other relations besides the subset relation. The basic set operations are union, intersection, difference, test for equality, and the operation 'is-an-element-of'. Table 3.1 contains definitions and the corresponding tools from the packages base and sets discussed below. In order to use the functions provided by the package sets, objects have to be defined as sets. All functions contained in base R can be applied to vectors or matrices. One can use the relations from Table 3.1 to state the following equations and properties, which are generally valid in set theory.

1. $A \cup \emptyset = A$, $A \cap \emptyset = \emptyset$;
2. $A \cup \Omega = \Omega$, $A \cap \Omega = A$;
3. $A \cup A^c = \Omega$, $A \cap A^c = \emptyset$;
4. $(A^c)^c = A$;
5. Commutative property: $A \cup B = B \cup A$, $A \cap B = B \cap A$;
6. Associative property: $(A \cup B) \cup C = A \cup (B \cup C)$, $(A \cap B) \cap C = A \cap (B \cap C)$;

Table 3.1 Definitions, relations and operations on sets

Notation	Definition	R base	sets package
$x \in A$	x is an element of A	`%in%,is.element()`	`set_contains_element(A,x),`
			`(x %e% A)`
$x \notin A$	x is not an element of A	`!(x %in% A)`	`!(x %e% A)`
$A \subseteq B$	Each element of A is an element of B	`A %in% B`	`set_is_subset(A, B)`
$A = B$	$A \supseteq B$ and $A \subseteq B$	`setequal(A,B)`	`set_is_equal(A, B)`
$\emptyset$	The empty set, {}	`x = c()`	`set()`
Ω	The Universe	`ls()`	
$A \cup B$	Union: $\{x \mid x \in A \text{ or } x \in B\}$	`union(A,B),`	`set_union(A,B), A	B`
$A \cap B$	Intersection: $\{x \mid x \in A \text{ and } x \in B\}$	`intersect(A,B)`	`set_intersection(A, B)`
	if $A \cap B = \emptyset$ then A and B are disjoint		`A & B`
$A \setminus B$	Set difference: $\{x \mid x \in A \text{ and } x \notin B\}$	`setdiff(A,B)`	`A - B`
$A \triangle B$	Symmetric difference: $(A \setminus B) \cup (B \setminus A)$		`set_symdiff(), %D%`
	$\{x \mid x \text{ either } x \in A \text{ or } x \in B\}$		
A^c	The complement of a set A: $\Omega \setminus A$		`set_complement(A, Ω)`
$\mathcal{P}(A)$	Power set: the set of all subsets of A		`set_power(A), 2^A`

7. Distributive property: $A \cup (B \cap C) = (A \cup B) \cap (A \cup C)$, $A \cap (B \cup C) = (A \cap B) \cup (A \cap C)$;
8. De Morgan's Law: $(A \cup B)^c = A^c \cap B^c$ and $(A \cap B)^c = A^c \cup B^c$, or, more generally, $(\cup_i A_i)^c = \cap_i A_i^c$ and $(\cap_i A_i)^c = \cup_i A_i^c$.

3.1.3 *Base Package*

The base package provides functions to perform most set operations, as shown in the second column of Table 3.1. The results are given as an output vector or list. Note that R is able to compare `numeric` and `character` elements. The output will be given as a `character` vector, as in line 3 below.

```
> set1 = c(1, 2)                 # numeric vector
> set2 = c("1", "2", 3)          # vector with strings
> setequal(set1, set2)           # sets are not equal
[1] FALSE
> is.element(set2, c(2, 1))      # 1, 2 are elements of 2nd set
[1]  TRUE  TRUE FALSE
> intersect(set1, set2)          # different element types
[1] "1" "2"
```

As there is no specific function in base package for the symmetric difference it can be obtained by combining base functions `union()` and `setdiff()` as:

```
> A =  1:4                          # {1, 2, 3, 4}
> B = -3:3                          # {-3, -2, -1, 0, 1, 2, 3}
> union(setdiff(A, B), setdiff(B, A))  # symmetric difference set
[1]   4 -3 -2 -1   0
```

The symmetric difference set is the union of the difference sets. In the example above, A and B have 1, 2, and 3 as their common elements. All other elements belong to the symmetric difference set.

When working with basic R objects like lists, vectors, arrays, or data-frames, using functions from the base package is appropriate. These functions, for example, union(), intersect(), setdiff() and setequal(), apply as.vector to the arguments. Applying operations on different types of sets, like a list and a vector in the following example, does not necessarily lead to a problem.

```
> setlist = list(3, 4)             # set of type list
> setvec1 = c(5, 6, 8, 20)         # set of type vector
> intersect(setlist, setvec1)      # no common elements
numeric(0)
> setvec2 = c("blue", "red", 3)    # set of type vector
> intersect(setlist, setvec2)      # common elements
[1] "3"
```

In the following example, the objects A and B are combined in the data frame AcB. The union of a data frame AcB and another object M returns a list of all elements.

```
> AcB = data.frame(A = 1:3, B = 5:7)
> M   = list(10, 15, 10)
> union(AcB, M)                    # union returns a list for data frames
[[1]]
[1] 1 2 3

[[2]]
[1] 5 6 7

[[3]]
[1] 10

[[4]]
[1] 15

> intersect(AcB, M)
list()
> DcE1 = data.frame(D = c(1, 3, 2), E = c(5, 6, 7))
> intersect(AcB, DcE1)    # should return both D and E
  E
1 5
2 6
3 7
> DcE2 = data.frame(D = c(1, 2, 3), E = c(5, 6, 7))
> intersect(AcB, DcE2)    # should return both D and E
  E
1 5
2 6
3 7
```

Using vectors as sets has some drawbacks when working with data frames, as shown for the intersections above. In the base package, the intersection of two data frames with a common element returns the empty set if the elements are ordered or defined differently, therefore the elements `c(1, 2, 3, 4)` and `c(1, 3, 2, 4)` as well as `c(1, 2, 3, 4)` and `1:4` are treated as different sets. When using the sets function `set()`, the order becomes unimportant.

3.1.4 *Sets* Package

The package `sets` was specifically created by David Meyer and others for applications concerning set theory. This package provides basic operations for ordinary sets and also for generalizations like fuzzy sets and multisets. The objects created with functions from this package, e.g. by using the function `set()`, can be viewed as real set objects, in contrast to vectors or lists, for example. This is visible in the output, since sets are denoted by curly brackets.

A data frame can be viewed as a nested set and should be created with several `set()` commands. Note that these functions in R require the `sets` package.

```
> require(sets)
> A = set(1, 2, 3)            # set A
> B = as.set(c(5, 6, 7))      # set B
> set(A, B)                   # set AcB from above
{{1, 2, 3}, {5, 6, 7}}
```

The `as.set()` function is used above to convert an array object into a set object. For objects of the class *set*, it is recommended to use the methods from the same package, like `set_union` and `set_intersection` or, more simply, the symbols & and |. In the following, some of these functions, as presented in Table 3.1, are used on two simple sets.

```
> A = set(1, 2, 3)            # set A
> B = set(5, 6, 7, "5")       # set B
> B                           # ordered and distinct
{"5", 5, 6, 7}
> A | B                       # union set
{"5", 1, 2, 3, 5, 6, 7}
> A & B                       # intersection set
{}
> A - B                       # set difference
{1, 2, 3}
> A %D% B                     # symmetric difference
{"5", 1, 2, 3, 5, 6, 7}
> summary(A %D% B)            # summary of the symmetric difference
A set with 7 elements.
> set_is_empty(A)             # check for empty set
[1] FALSE
```

Besides the functions in Table 3.1, the basic predicate functions ==, != , <, <=, defined for equality and subset, can be used intuitively for the set objects. For vectors or lists, however, these functions are executed element by element, so the objects must have the same length.

```
> A = set(1, 2, 3); B = set(5, 6, 7, "5")
> D = set(4, 5, 6, 7, "5")  # set D
> B <= D                        # check if B is a subset of D
[1]  TRUE
> B != A                        # check if B is unequal to A
[1]  TRUE
> set_similarity(B, D)          # fraction of elements in intersection
[1]  0.8
```

Where the set_similarity function computes the fraction of the number of the elements in the intersection of two sets over the number of the elements in the union. In computational statistics, one often needs to work with sets and compute such properties as the mean and median, see Sect. 5.1.5. Such statistics can be calculated for set objects similarly to other R objects. Applying the functions sum(), mean() and median() to a set, R will try to convert the set to a numeric vector, e.g. 5 defined as a character is converted to a numeric 5 in the example below.

```
> A = set(1, 2, 3); B = set(5, 6, 7, "5")
> A + B                         # union of A and B
{ "5", 1, 2, 3, 5, 6, 7}
> sum(c("5", 1, 2, 3, 5, 6, 7))
Error in sum(c("5", 1, 2, 3, 5, 6, 7)):
    invalid 'type' (character) of argument
> sum(A + B)                    # sum of union set A and B
[1] 29                          # "5" becomes numeric
> A * B                         # Cartesian product
{(1, 5), (1, 6), (1, 7), (1, "5"), (2, 5), (2, 6), (2, 7), (2, "5"),
 (3, 5), (3, 6), (3, 7), (3, "5")}
```

Furthermore, in the sets package, the calculation of the closure and reduction of sets is implemented by means of the function closure().

```
> D = set(set(1), set(2), set(3)); D
{{1}, {2}, {3}}
> closure(D)                    # set of all subsets
{{1}, {2}, {3}, {1, 2}, {1, 3}, {2, 3}, {1, 2, 3}}
```

3.1.5 Generalised Sets

In contrast to ordinary sets generalised sets allow keeping, which have their elements in a sorted and distinct form, a generalised set keeps every element, even if there are redundant elements, but still in a sorted way. Generalised sets allow keeping more information or characteristics of a set and include two special cases: fuzzy sets and multisets. Every generalised set can be created using the gset() function and all methods in this regard begin with the prefix gset_. Before constructing a generalised set, it is important to think about its characteristics, like the membership of an element, which differ for fuzzy sets and multisets.

Membership is described by a function f that maps each element of a set A to a membership number:

- For ordinary sets, each element is either in the set or not, i.e. $f : A \rightarrow \{0, 1\}$;
- For fuzzy sets, the membership function maps into the unit interval, $f : A \rightarrow [0, 1]$;
- For multisets, $f : A \rightarrow \mathbb{N}$.

Multisets allow each element to appear more than once, so that in statistics, multisets occur as frequency tables. Since in base R there is no support for multisets, the sets package is a good solution. In the example below, the set object A has four distinct elements and each element has a certain membership value. The absolute *cardinality* of a set can be obtained by the function gset_cardinality(), i.e. the number of elements in a set.

```
> require(sets)
> ms1 = gset(c("red", rep("blue", 3)))  # multiset
> ms1                                     # repeated elements retained
{"blue", "blue", "blue", "red"}
> gset_cardinality(ms1)                   # number of elements
[1] 4                                     # cardinality of ms1
> fs1 = gset(c(1, 2, 3),                  # fuzzy set
+    membership = c(0.2, 0.6, 0.9));
> fs1
{1 [0.2], 2 [0.6], 3 [0.9]}
> plot(fs1)                               # left plot in Fig.\,3.1
> B = c("x", "y", "z", "z", "z", "x")     # create multiset from R object
> table(B)
B
x y z
2 1 3
> ms2 = as.gset(B); ms2                   # converts vector to the set
{"x" [2], "y" [1], "z" [3]}
> gset_cardinality(ms2)                   # cardinality of multi2
[1] 6
> ms3 = gset(c('x', 'y' , 'z'),           # create multiset via gset
+    membership = c(2, 1, 3)); ms3
{"x" [2], "y" [1], "z" [3]}
> gset_cardinality(ms3)
[1] 6
> plot(ms3, col = 'lightblue')            # right plot in Fig.\,3.1
```

By employing the repeat function rep(x, times) with times = 2, the membership is doubled.

```
> ms4 = rep(ms3, times = 2)
> ms4
{"x" [4], "y" [2], "z" [6]}
> gset_cardinality(ms4)
[1] 12
```

The function set_combn(set, length) from the sets package creates a set with subsets of the specified length: it consists of all combinations of the elements in the specified set (Fig. 3.1).

When the same function is applied to all factorial combinations of two sets, e.g. the function set_outer(set1, set2, operation) applies a binary operator like sum or product to all elements of two sets. It applies the operation to all pairs of elements specified in sets 1 and 2 and returns a matrix of dimension length(set1) times length(set2). outer can be also used for vectors and matrices in R.

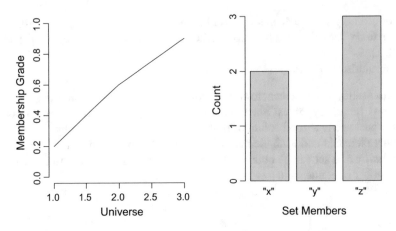

Fig. 3.1 R plot of a fuzzy set (*left*) and a multiset (*right*). Q BCS_FuzzyMultiSets

```
> set_combn(set(2, 4, 6, 8, 10),           # all subsets
+    length = 2)                            # of length 2
{{2, 4}, {2, 6}, {2, 8}, {2, 10}, {4, 6}, {4, 8}, {4, 10}, {6, 8},
 {6, 10}, {8, 10}}
> set_outer(set(3, 4),
+    set(10, 50, 100), "*")                 #outer product with sets
   10  50 100
3  30 150 300
4  40 200 400
> outer(c(3, 4), c(10, 50, 100), "*")       #outer product with vectors
     [,1] [,2] [,3]
[1,]   30  150  300
[2,]   40  200  400
```

Users of base R can get wrong or confusing results when applying basic set operations like union and intersection. Indexable structures, like lists and vectors, are interpreted as sets. For set theoretical applications, this imitation has not been sufficiently elaborated: basic operations such as the Cartesian product and power set are missing. The base package in R performs a type conversion via match(), which might in some cases lead to wrong results. In most cases it makes no difference whether one uses a = 2 or a = 2L, where the latter defines a directly as an integer by the suffix L. But to save memory in computationally extensive codes, it is useful to define a directly as having integer type.

```
> y = (1:100) * 1                           # option 1 to define vector y
> typeof(y)
[1] "double"
> object.size(y)                            # memory used by this object
840 bytes
> yL = (1:100) * 1L                         # option 2 to define vector y
> typeof(yL)
[1] "integer"
> object.size(yL)                           # memory used by this object
440 bytes
```

Fig. 3.2 Inheritance of object classes `set`, `gset` and `cset` from the `sets` package. All operation names can be combined with the corresponding prefix `set_`, `gset_` and `cset_`

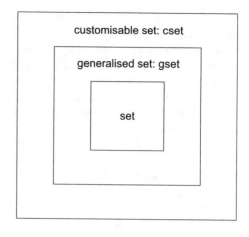

If one tries to check its code for constants not defined as integers, the `match()` function will not distinguish between `1` and `1L`.

The `sets` package avoids such steps by the use of set classes for ordinary, general, and customised sets, as presented in Fig. 3.2. Customised sets are an extension of generalized sets and are implemented in R via the function `cset`. With the help of customisable sets, one is able to define how elements in the sets are matched through the argument `matchfun`.

```
> setA   = set(as.numeric(1))        # set with numeric 1
> 1L %e% setA                        # 1L is not an element of A
[1] FALSE
> csetA = cset(as.numeric(1),        # cset with match function
+   matchfun = match)
> 1L %e% csetA                       # 1L is now an element of A
[1] TRUE
```

The basic R function `match` considers the integer one `1L` to be the same as the numeric `1`. With the help of customisable sets, users of R are able to specify which elements are considered to be the same. This is very useful for data management.

3.2 Probabilistic Experiments with Finite Sample Spaces

When working with data that is subject to random variation, the theory behind this probabilistic situation becomes important. There are two types of experiments: deterministic and random. We will focus here on nondeterministic processes with a finite number of possible outcomes.

A random trial (or experiment) yields one of the distinct outcomes that altogether form the sample or event space Ω. All possible outcomes constitute the universal event. Subsets of Ω are called events, e.g. Ω itself is the universal event. Examples of experiments include rolling a die with the sample space $\Omega =$

{{1}, {2}, {3}, {4}, {5}, {6}}, another is tossing a coin with only two possible outcomes: heads (H) and tails (T).

A combination of several rolls of a die or tosses of a coin leads to more possible results, such as tossing a coin twice, with the sample space $\Omega = \{\{H, H\}, \{H, T\}, \{T, H\}, \{T, T\}\}$. Generally, the combination of several different experiments yields a sample space with all possible combinations of the single events. If, for instance, one needs two coins to fall on the same side, then the favored event is a set of two elements: $\{H, H\}$ and $\{T, T\}$.

The `prob` package, which will be used in the following, has been developed by G. Jay Kerns specifically for probabilistic experiments. It provides methods for elementary probability calculation on finite sample spaces, including counting tools, defining probability spaces discussed later, performing set algebra, and calculating probabilities.

The situation of tossing a coin twice is considered in the following code, for which the package `prob` is needed. The functions used will be explained shortly.

```
> require(prob)
> ev = tosscoin(2)          # sample space for 2 coin-tosses
> probspace(ev)             # probabilities for events
  toss1 toss2 probs
1    H     H   0.25
2    T     H   0.25
3    H     T   0.25
4    T     T   0.25
```

The interesting information is how likely an event is. Each event has a probability assigned and this probability is included as the last column of the R output in the example above. The values quantify our chances of observing the corresponding outcome for the outcome of tossing a coin twice.

Comparable to the set theory in Sect. 3.1, one can apply operations like union or intersection to events. The event probability follows the axioms of probability, which are shortly summarised in the following.

- $P(\cdot)$ is a probability function that assigns to each event A in the sample space a real number $P(A)$, which lies between zero and one. $P(A)$ is the probability that the event A occurs. The probability of the whole sample space is equal to one, which means that it occurs with certainty.
- $P(A \cup B) = P(A) + P(B)$ if A and B are disjoint. In general,
  $P(A \cup B) = P(A) + P(B) - P(A \cap B)$,
  $P(\Omega) = 1$ and $P(\emptyset) = 0$.

The probability of the complementary event and that of the difference between two sets are given by

- $P(A^c) = 1 - P(A)$;
- $P(A \setminus B) = P(A) - P(A \cap B)$.

3.2.1   R Functionality

The following functions, which generate common and elementary experiments, can be used to set up a sample space.

- `urnsamples(x, size, replace = FALSE, ordered = FALSE, ...)`,
- `tosscoin(ncoins, makespace = FALSE)`,
- `rolldie(ndies, nsides = 6, makespace = FALSE)`,
- `cards(jokers = FALSE, makespace = FALSE)`,
- `roulette(european = FALSE, makespace = FALSE)`.

If the argument `makespace` is set TRUE, the resulting data frame has an additional column showing the (equal) probability of each single event. In the simplest case, the probability of an event can be computed as the relative frequency. Some methods for working with probabilities and random samples from the `prob` and the `base` packages are the following.

- `probspace(outcomes, probs)` forms a probability space,
- `prob(prspace, event = NULL)` gives the probability of an event as its relative frequency,
- `factorial(n)` is the mathematical operation $n!$ for a non-negative integer n,
- `choose(n, k)` gives the binomial coefficient $\binom{n}{k} = \frac{n!}{k!(n-k)!}$.

```
> require(prob)
> ev = urnsamples(c("bus", "car", "bike", "train"),
+     size    = 2,
+     ordered = TRUE)
> probspace(ev)                              # probability space
      X1    X2       probs
1    bus   car 0.08333333
2    car   bus 0.08333333
3    bus  bike 0.08333333
4   bike   bus 0.08333333
5    bus train 0.08333333
6  train   bus 0.08333333
7    car  bike 0.08333333
8   bike   car 0.08333333
9    car train 0.08333333
10 train   car 0.08333333
11  bike train 0.08333333
12 train  bike 0.08333333
> Prob(probspace(ev), X2 == "bike")    # 3 of 12 cases = 1 / 4
[1] 0.25
> factorial(3)                         # 3 * 2 * 1
[1] 6
> choose(n = 10, k = 2)                # 10! / (2! * 8!) = 10 * 9 / 2
[1] 45
```

3.2.2   Sample Space and Sampling from Urns

In R, the sample spaces can be represented by data frames or lists and may contain empirical or simulated data. Random samples, including sampling from urns, can be drawn from a set with the R base method `sample()`. The sample size can be

Table 3.2 Number of all possible samples of size k from a set of n objects. The sampling method is specified by replacement and order

	Ordered	Unordered
With replacement	n^k	$\frac{(n+k-1)!}{k!(n-1)!}$
Without replacement	$\frac{n!}{(n-k)!}$	$\binom{n}{k}=\frac{n!}{k!(n-k)!}$

chosen as the second argument in the function and the type of sampling can be either with or without replacement:

- sampling with replacement:
  ```
  sample(x, size = n, replace = TRUE, prob = NULL),
  ```
- sampling without replacement:
  ```
  sample(x, n).
  ```

In general, there are four types of sampling, regarding replacement and order, which are briefly presented in the following. The calculation rules for the number of possible draws for a sample depend on the assumptions about the particular situation. All four cases are outlined in Table 3.2.

In R the function nsamp is able to calculate the possible numbers of samples drawn from an urn. The following code shows how all four cases from Table 3.2 are applied when $n = 10$ and $k = 2$.

```
> require(prob)
> nsamp(10, 2, replace = TRUE, ordered = TRUE)     # 10^2
[1] 100
> nsamp(10, 2, replace = TRUE, ordered = FALSE)    # 11! / (2! * 9!)
[1] 55
> nsamp(10, 2, replace = FALSE, ordered= TRUE)     # 10! / 8!
[1] 90
> nsamp(10, 2, replace = FALSE, ordered = FALSE)   # 10! / (2! * 8!)
[1] 45
```

Ordered Sample

For several applications, the order of k experimental outcomes is decisive. Consider, for example, the random selection of natural numbers. For the random selection of a telephone number, both the replacement and the order of the digits are important.

The method urnsample() from the prob package yields all possible samples according to the sampling method. Consider the next example, where three elements are taken from an urn of eight elements. Sampling with replacement is conducted first, followed by sampling without replacement for comparison. Clearly the number of samples is smaller if we do not replace the elements. This number can also be computed with the counting tool nsamp() introduced in the last section.

```
> require(prob)
> urn1 = urnsamples(x = 1:3,    # all elements
+    size      = 2,             # num of selected elements
+    replace  = TRUE,           # with replacement
+    ordered  = TRUE)           # ordered
> urn1                          # all possible draws
   X1 X2
1   1  1
2   2  1
3   3  1
4   1  2
5   2  2
6   3  2
7   1  3
8   2  3
9   3  3
> dim(urn1)                     # dimension of the matrix
[1] 9 2
> urn2 = urnsamples(x = 1:3,
+    size      = 2,
+    replace = FALSE,           # without replacement
+    ordered = TRUE)            # ordered
> dim(urn2)                     # dimension of the matrix
[1] 6 2
```

Unordered Sample

In the simple case of drawing balls from an urn, the order in which the balls are drawn is rarely relevant. For a lottery, for example, it is only relevant whether a certain number is included in the winning sample or not. When conducting a survey or selecting participants, the order of the selection is generally irrelevant. Having created the sample space, a sample can be drawn, which leaves the question about the replacement. The researcher has to decide what fits best in this situation.

Note that in an unordered sample without replacement, the number of possible samples is given by the binomial coefficient. Using the formula from Table 3.2, the sample size can be checked and the probability of drawing a certain sample can be calculated.

```
> require(prob)
> urn3 = urnsamples(x = 1:3,
+    size      = 2,
+    replace = TRUE,            # with replacement
+    ordered = FALSE)           # not ordered
> dim(urn3)                     # dimensions of the matrix
[1] 6 2
> urn4 = urnsamples(x = 1:3,
+    size      = 2,
+    replace = FALSE,           # without replacement
+    ordered = FALSE)           # not ordered
> urn4                          # all possible draws
   X1 X2
1   1  2
2   1  3
3   2  3
> probspace(urn4)              # probability space
   X1 X2      probs
1   1  2 0.3333333
2   1  3 0.3333333
3   2  3 0.3333333
```

The probability of obtaining a certain pair of values is one over the number of possible pairs. For the case without replacement and ignoring order, each sample has the probability $1/3 \approx 0.3333$. This number together with all 3 possible samples is given when applying the method `probspace()` to `urn4`.

Beside these simple experiments, it is also useful to know that the number of subsets of a set of n elements is 2^n. Furthermore, there are $n!$ possible ways of choosing all n elements and rearranging them. This is the same thing as the number of permutations of n elements. In case the sample size is the same as the number of elements and `replace = FALSE`, the sampling can be seen as a random permutation. If the sample space consists of all combinations of a number of factors, the function `expand.grid()` from the `base` package can be used to generate a data frame containing all combinations of these factors. The example below shows all combinations of two variables specifying colour and number.

```
> expand.grid(colour = c("red", "blue", "yellow"), nr = 1:2)
    colour nr
1      red  1
2     blue  1
3   yellow  1
4      red  2
5     blue  2
6   yellow  2
```

There are several ways to sample from a population. It matters for the number of possible samples whether one arranges the elements or selects a subset from the population. All different possibilities are illustrated in Fig. 3.3.

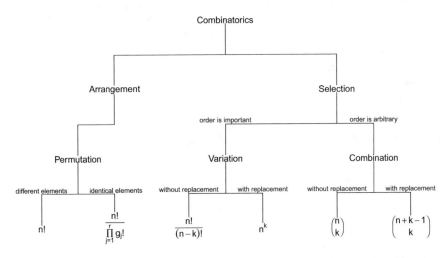

Fig. 3.3 The possible sample numbers for an urn model with n elements. For samples with assembled elements with r groups for identical elements g_j or k out of n selected different elements.
🔍 BCS_SamplesDiagram

3.2.3 Sampling Procedure

The examples above are very specific and restricted to a particular sample space. Now we can address sampling from a more general perspective. Again, some random selection mechanism is involved: the theory behind this is called probabilistic sampling. Specific types of sampling are: simple random sampling, the equal probability selection method, probability-proportional-to-size, and systematic sampling. Details can be found in Babbie (2013). In real applications, each member of a population can have different characteristics, i.e. the population is heterogeneous, and one needs a sample large enough to study the characteristics of the whole population. The idea is to find a sample which describes the population well. Yet, there is always a risk of biased samples if the sampling method is not adequate, that is to say, if the set of selected members is not representative of the population.

In the following example, it is assumed that a population consists of women and men in a ratio of 1 : 1. In order to test this assumption about the ratio, a sample is drawn.

```
> # set.seed(18)                        # set the seed, see Chap.\,9
> popul = data.frame(
+   gender = rep(c("f", "m"), each = 500),
+   grade  = sample(1:10, 1000, replace = TRUE))
> head(popul)                           # first 6 rows of matrix
  gender grade
1      f     9
2      f     8
3      f    10
4      f     1
5      f     1
6      f     6
> table(popul[, 1])                     # true proportion
  f   m
500 500
> table(sample(popul[, 1], 10))  # draw sample of 10
f m
3 7
```

In this example, a simple random sample was drawn, which was too small to capture the true ratio. For more sophisticated sampling methods in R, the package `sampling` can be used. It contains methods for stratified sampling, which divides the population into subgroups and samples. The corresponding R function is `strata()`. Its argument, `stratanames`, specifies the variable that is used to identify the subgroups.

```
> require(sampling)
> strata(data, stratanames = NULL, size,
+   method = c("srswor", "srswr", "poisson", "systematic"),
+   pik, description = FALSE)
```

The two methods, `srswor` and `srswr`, denote simple random sampling without and with replacement, respectively. In the example below, a sample of six persons each is taken from the female and male students without replacement.

The function `getdata()` extracts data from a dataset according to a vector of selected units or a sample data frame. Here, we use the sample data frame created by the function `strata()` to extract the grades for the sample students from our dataset.

A simple tool of analysis is the function `aggregate()`, which is used to calculate summary statistics for subsets of data. It is applied below to calculate the mean of the grades in the sample for each gender. Note, that the subsets need to be given as a list.

```
> require(sampling)
> st    = strata(popul,
+     stratanames = "gender",      # take 6 samples of each gender
+     size        = c(6, 6),
+     method      = "srswor")
> dataX = getdata(popul, m = st) # extract the sample
> dataX
      grade gender ID_unit  Prob Stratum
98        8      f      98 0.012       1
114       5      f     114 0.012       1
288       1      f     288 0.012       1
392       5      f     392 0.012       1
411       7      f     411 0.012       1
421       6      f     421 0.012       1
532       2      m     532 0.012       2
619       9      m     619 0.012       2
667       7      m     667 0.012       2
771       3      m     771 0.012       2
952       1      m     952 0.012       2
968       3      m     968 0.012       2
> aggregate(data$grade,          # mean grade by gender
+     by  = list(data$gender),
+     FUN = mean)
  Group.1        x
1       f 5.333333
2\,m 4.166667
```

To test whether these results support our expectations of equal grades for each gender, we would need some functions for statistical testing discussed in Sect. 5.2.2. Applying a t-test, we would find that the results are indeed supportive.

3.2.4 Random Variables

The outcomes of a probabilistic experiment might be described by a random variable X. All possible events ω are elements of Ω, the event space. These outcomes of a probabilistic experiment are associated with distinct values x_j of X, for $j = \{1, \ldots, k\}$.

Definition 3.1 A real valued *random variable* (rv) X on the probability space $(\Omega, \mathcal{F}, P)$, is a real valued function $X(\omega)$ defined on Ω, such that for every Borel subset $\mathcal{B}$ of the real numbers

$$\{\omega : X(\omega) \in \mathcal{B}\} \in \mathcal{F}.$$

The probability function P assigns a probability to each event, for a detailed discussion, see Ash (2008).

For the probabilistic experiment of tossing a fair coin $\Omega = \{H, T\}$, the rv X is defined as follows:

$$X = \begin{cases} 1, & \text{if head } H \text{ shows up}, \\ 0, & \text{if tail } T \text{ shows up}. \end{cases}$$

There are two types of rvs: discrete and continuous. This distinction is very important for their analysis.

Definition 3.2 An rv X is said to be *discrete* if the possible distinct values x_j of X are either countably infinite or finite.

The distribution of a discrete rv is described by its probability mass function $f(x_j)$ and the cumulative distribution function $F(x_j)$:

Definition 3.3 The probability mass function (pdf) of a discrete rv X is a function that returns the probability, that an rv X is exactly equals to some value

$$f(x_j) = P(X = x_j).$$

Definition 3.4 The cumulative distribution function (cdf) is defined for ordinally scaled variables (variables with the natural order) and returns the probability, that an rv X is smaller or equal to some value:

$$F(x_j) = P(X \leq x_j).$$

The outcomes of tossing a fair coin can be mapped by a discrete rv with finite distinct values. Drawing randomly a person and its number of descendants can be described by a discrete rv with countably infinite distinct values.

An rv X has an expectation $E X$ and a variance $Var X$ (also called the first moment and the second central moment of X, respectively). The definition of these moments differs for discrete and continuous rvs.

Definition 3.5 Let X be a discrete rv with distinct values $\{x_1, \ldots, x_k\}$ and probability function $P(X = x_j) \in [0, 1]$ for $j \in \{1, \ldots, k\}$. Then the *expectation* (*expected value*) of X is defined to be

$$E X = \sum_{j=1}^{k} x_j P(X = x_j). \tag{3.1}$$

For infinitely many possible outcomes, the finite sum becomes an infinite sum. The expectation is not defined for every rv. An example for continuous rvs is the Cauchy distribution, introduced in Sect. 4.5.2.

Definition 3.6 Let X be a discrete rv with distinct values $\{x_1, \ldots, x_k\}$ and probability function $P(X = x_j) \in [0, 1]$ for $j \in \{1, \ldots, k\}$. Then the *variance* of X is defined to be

$$\text{Var } X = E(X - E\, X)^2 = \sum_{j=1}^{k} (x_j - E\, X)^2 P(X = x_j). \qquad (3.2)$$

As for the expectation, the variance is not defined for every rv. The variance measures the expected dispersion of an rv around its expected value. Deterministic variables have a variance equal to zero.

Definition 3.7 An rv X is said to be *continuous* if the possible distinct values x_j are uncountably infinite.

For a continuous rv, the probability density function (pdf) describes its distribution (see Definition 4.2). Selecting randomly a person and its weight is a typical example of a probabilistic experiment which can be described by a continuous rv.

In the following, the most prominent discrete rvs and their probability mass functions are introduced. Continuous rvs and their properties are covered in Chap. 4.

3.3 Binomial Distribution

One of the basic probability distributions is the binomial. Examples of this distribution can be observed in daily life: whether we are tossing a coin to obtain heads or tails, or trying to score a goal in a football game, we are dealing with a binomial distribution.

3.3.1 Bernoulli Random Variables

A *Bernoulli experiment* is a random experiment with two outcomes: success or failure. Let p denote the probability of the success of each trial and let the rv X be equal to 1 if the outcome is a success and 0 if a failure. Then the probability mass function of X is

$$P(X = 0) = 1 - p,$$
$$P(X = 1) = p$$

and the rv X is said to have a *Bernoulli distribution*. The expected value and variance of a Bernoulli rv are $E\, X = p$ and $\text{Var } X = p(1 - p)$.

To derive these results, just apply (3.1) and (3.2). The expectation is then derived as follows:

$$\mathsf{E}X = P(X = 0) \cdot 0 + P(X = 1) \cdot 1 = (1 - p) \cdot 0 + p = p.$$

For the variance, the derivation looks as follows:

$$\begin{aligned}
\mathsf{Var}\,X &= P(X = 0)(0 - p)^2 + P(X = 1)(1 - p)^2, \\
&= (1 - p)p^2 + p(1 - p)^2, \\
&= p^2 - p^3 + p - 2p^2 + p^3, \\
&= p(1 - p).
\end{aligned}$$

Example 3.1 Consider a box containing two red marbles and eight blue marbles. Let $X = 1$ if the drawn marble is red and 0 otherwise. The probability of randomly selecting one red marble and the expectation of X at one try is $\mathsf{E}X = P(X = 1) = 1/5 = 0.2$. The variance of X is $\mathsf{Var}\,X = 1/5(1 - 1/5) = 4/25$.

3.3.2 Binomial Distribution

A sequence of Bernoulli experiments performed n times is called a *binomial experiment* and satisfies the following requirements:

1. there are only two possible outcomes for each trial;
2. all the trials are independent of each other;
3. the probability of each outcome remains constant;
4. the number of trials is fixed.

The following example illustrates a binomial experiment. When tossing a fair coin three times, eight different possible results may occur, each with equal probability 1/8. Let X be the rv denoting the number of heads obtained in these three tosses. The sample space Ω and possible values for X are listed in Table 3.3. One is often interested in the total number of successes and the corresponding probabilities, instead of the outcomes themselves or the order in which they occur. For the above case, it follows that

$$P(X = 0) = \frac{|\{TTT\}|}{|\Omega|} = 1/8; \qquad P(X = 2) = \frac{|\{HHT, HTH, THH\}|}{|\Omega|} = 3/8;$$

Table 3.3 Sample space and X for tossing a coin three times

Outcome	HHH	HHT	HTH	THH	HTT	THT	TTH	TTT
Value of X	3	2	2	2	1	1	1	0

$$P(X = 3) = \frac{|\{HHH\}|}{|\Omega|} = 1/8; \quad P(X = 1) = \frac{|\{HTT, THT, TTH\}|}{|\Omega|} = 3/8.$$

The same results can be computed from the binomial mass function below by setting $n = 3$ and $p = 1/2$.

Definition 3.8 The *binomial distribution* is the distribution of an rv X for which

$$P(X = x) = \binom{n}{x} p^x (1 - p)^{n-x}, \quad x \in \{0, 1, 2, \ldots, n\}, \quad (3.3)$$

where n is the number of all trials, x is the number of successful outcomes, p is the probability of success, $\binom{n}{x}$ is the number of possibilities of n outcomes' leading to x successes and $n - x$ failures.

The binomial distribution can be used to define the probability of obtaining exactly x successes in a sequence of n independent trials. We will denote binomial distributions by $B(x; n, p)$ For the example above, the rv X follows the binomial distribution $B(x; 3, 0.5)$, or $X \sim B(3, 0.5)$. The expectation of a binomial rv is $E\,X = \mu = np$, which is the expected number of successes x in n trials. The variance is $\text{Var}\,X = \sigma^2 = np(1 - p)$.

Example 3.2 Continuing with marbles, we randomly draw ten marbles one at a time, while putting it back each time before drawing again. What is the probability of drawing exactly two red marbles?

Here, the number of draws $n = 10$ and we define getting a red marble as a success with $p = 0.2$ and $x = 2$. Hence $X \sim B(10, 0.2)$ and

$$P(X = x) = \binom{10}{x} 0.2^x (1 - 0.2)^{10-x}, \quad x = 0, 1, 2, \ldots, 10$$

and $\mu = 2$, $\sigma^2 = 1.6$. For $x = 2$ we obtain $P(X = 2) = 0.30199$.

In R, the command dbinom() creates a probability mass function (3.3). The desired probability of the example above is obtained as follows.

```
> dbinom(x = 2,          # probability mass function at x = 2
+    size = 10,          # number of trials
+    prob = 0.2)         # probability of success
[1] 0.30199
```

Furthermore, one can use dbinom() to calculate the probability of each outcome (Fig. 3.4).

Fig. 3.4 Probability mass function of the binomial distribution with number of trials $n = 10$ and probability of success $p = 0.2$.
Q BCS_Binhist

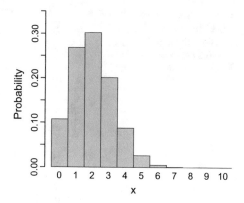

```
> dbinom (x = 0:10,      # probability at 0, 1, ..., 10
+     size = 10,         # number of trials
+     prob = 0.2)        # probability of success
 [1]  0.107374  0.268435  0.301989  0.201326  0.088080
 [6]  0.026424  0.005505  0.000786  0.000074  0.000004
[11]  0.000000
```

The Cumulative Distribution Function

The cdf of a binomial distribution for discrete variables is defined as

$$F_X(x) = P(X \le x) = \sum_{i=0}^{x} \binom{n}{i} p^i (1 - p)^{n-i}.$$

It is implemented in R by pbinom().

Example 3.3 Continuing Example 3.2, consider the probability of drawing two or less red marbles. Let $n = 10$, $p = 0.2$ and $x = 2$, then

$$F_X(2) = P(X \le 2) = P(X = 0) + P(X = 1) + P(X = 2)$$

$$= \sum_{i=0}^{2} \binom{10}{i} 0.2^i (1 - 0.2)^{10-i} = 0.6777995.$$

```
> pbinom (x = 2, size = 10, prob = 0.2)
[1] 0.6777995
```

Equivalently, the result can be obtained using dbinom().

```
> sum (dbinom (x = 0:2, size = 10, prob = 0.2))
[1] 0.6777995
```

Fig. 3.5 The binomial
cumulative distribution
function with $n = 10$,
$p = 0.2$ and $p = 0.6$.
⊘ BCS_Bincdf

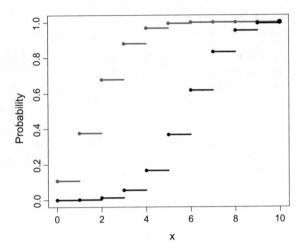

The probability that three or four red marbles are drawn is given by (Fig. 3.5).

```
> pbinom(4, size = 10, prob = 0.2) - pbinom(2, size = 10, prob = 0.2)
[1]   0.289407      # F(4) - F(2)

> qbinom(p = 0.95, size = 10, prob = 0.2)
[1] 4
```

3.3.3 *Properties*

A binomial distribution is symmetric if the probability of each trial is $p = 0.5$. Given
two binomial rvs $X \sim B(n, p)$ and $Y \sim B(m, p)$, the sum of the two rvs also follows
a binomial distribution $X + Y \sim B(n + m, p)$ with expectation $(m + n)p$. Intuitively,
n independent Bernoulli experiments and another m independent Bernoulli experiments again follow a binomial distribution: that of $(n + m)$ independent Bernoulli
experiments. The De Moivre–Laplace CLT ensures that the binomial rv converges
in distribution to the normal distribution (see Sect. 4.3), i.e.

$$Z = \frac{X - np}{\sqrt{np(1 - p)}} \xrightarrow{\mathcal{L}} N(0, 1)$$

as $n \to \infty$. Different textbooks recommend various rules for values of n and p that
will make the normal distribution (see Sect. 4.3) a good approximation for the binomial distribution (see Figs. 3.6 and 3.7). Values fulfilling $np > 5$ and $n(1 - p) > 5$
already produce satisfying results (Brown et al. (2001)). Since the binomial distribution is a discrete distribution and the normal distribution is continuous, the correction

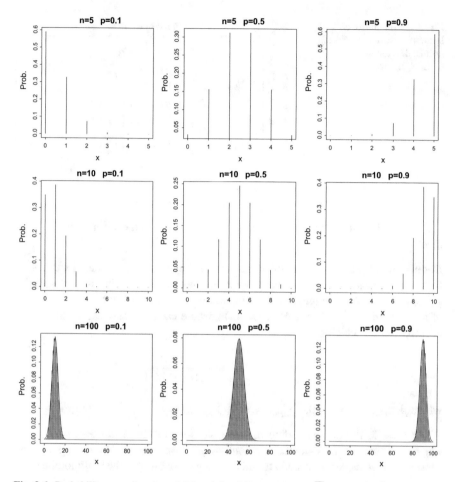

Fig. 3.6 Probability mass function of $B(n, p)$ for different n and p. ⌕ BCS_Binpdf

for continuity requires adding or subtracting 0.5 from the values of the discrete binomial rv. Furthermore, the binomial distribution can approach other distributions in the limit. If $n \to \infty$ and $p \to 0$ with finite np, the limit of the binomial distribution is the Poisson distribution, see Sect. 3.6. A hypergeometric distribution can also be obtained from the binomial distribution under certain conditions, see Sect. 3.5.

3.4 Multinomial Distribution

A binomial experiment always generates two possible outcomes (success or failure) at each trial. When tossing a die, the probability of getting the number six is $1/6$, and there is a $5/6$ chance of rolling any other number. But if we toss several dice

Fig. 3.7 Probability of
binomial distribution versus
normal distribution.
Q BCS_Binnorm

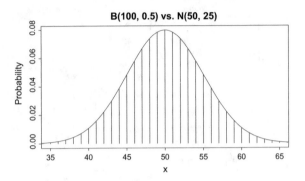

together each time, what is the probability of getting only a certain number for all
dice?

Definition 3.9 Suppose a random experiment is independently repeated n times, so
that it returns each time one of the fixed k possible outcomes with the probabilities
$p_1, p_2, \ldots, p_k$. An example of the *multinomial distribution* arises as the distribution
of a vector of rvs $X = (X_1, X_2, \ldots, X_k)^\top$ where each X_i denotes the number of
occurrences for which

$$P(X_1 = x_1, X_2 = x_2, \ldots, X_k = x_k) = \frac{n!}{x_1! x_2! \cdots x_k!} p_1^{x_1} p_2^{x_2} \cdots p_k^{x_k}, \qquad (3.4)$$

where $p_1, p_2, \ldots, p_k > 0$, $\sum_{i=1}^{k} p_i = 1$, $\sum_{i=1}^{k} x_i = n$, and x_i is nonnegative.

When $k = 3$, the corresponding distribution is called the *trinomial distribution*.
For $k = 2$, we get the binomial distribution discussed above. The example below
illustrates how to use the formula to calculate the probability in the multinomial case.

Example 3.4 Suppose we had a box with two red, three green, and five blue marbles. We randomly draw three marbles with replacement. What is the probability of
drawing one marble of each colour?

Here the realizations of the rv are $x_1 = 1$, $x_2 = 1$, $x_3 = 1$, and the corresponding
probabilities are $p_1 = 0.2$, $p_2 = 0.3$, $p_3 = 0.5$. Therefore according to (3.4), the
desired probability is $P(X_1 = 1, X_2 = 1, X_3 = 1) = \frac{3!}{1! \, 1! \, 1!} 0.2^1 0.3^1 0.5^1 = 0.18$.

In R, this can be calculated as follows.

```
> dmultinom(x = c(1, 1, 1),      # set success probabilities
+      prob = c(0.2, 0.3, 0.5)) # set values of multinomial rvs
[1] 0.18
```

Each specific rv X_i, for $i = 1, 2, \ldots, k$, follows a binomial distribution, thus the
expectation and the variance are $\mathsf{E}\, X_i = np_i$ and $\mathsf{Var}\, X_i = np_i(1 - p_i)$, respectively.

3.5 Hypergeometric Distribution

In the typical '6 from 49' lottery, 6 numbers from 1 to 49 are chosen without replacement. Every time one number is drawn, the chances of the remaining numbers to be chosen will change. This is an example of a *hypergeometric experiment*, which satisfies the following requirements:

1. a sample is randomly selected *without replacement* from a population;
2. each element of the population is from one of two different groups which can also be defined as success and failure.

Because the sample is drawn without replacement, the trials in the hypergeometric experiment are not independent and the probability of each success in turn keep changing. This differs from the binomial and multinomial distributions.

Definition 3.10 An rv X from a hypergeometric experiment follows the *hypergeometric distribution* $H(n, M, N)$, which has the probability function

$$P(X = x) = \frac{\binom{M}{x}\binom{N-M}{n-x}}{\binom{N}{n}}, \qquad x = 0, 1, ..., \min\{M, n\}, \qquad (3.5)$$

where N is the size of the population, n is the size of the sample, M is the number of successes in the population and x is the number of successes in the sample.

In (3.5), the probability of exactly x successes in n trials of a hypergeometric experiment is given. The following example illustrates this distribution.

Example 3.5 Having a box with 20 marbles, including 10 red and 10 blue marbles, we randomly select 6 marbles without replacement. The probability of getting two red marbles can be calculated using the $H(6, 10, 20)$ distribution. Here the experiment consists of 6 trials, so $n = 6$, and there are 20 marbles in the box, so $N = 20$. Now, $M = 10$, since there are 10 red marbles inside, of which two should be selected, so $x = 2$. Then

$$P(X = 2) = \frac{\binom{10}{2}\binom{20-10}{6-2}}{\binom{20}{6}} = 0.2438.$$

This example has a straightforward solution in R.

```
> dhyper (x = 2,   # number of successes in sample
+     m = 10,      # number of successes in population
+     n = 10,      # number of fails in population
+     k = n)       # sample size
[1] 0.243808
```

The binomial distribution is a limiting form of the hypergeometric distribution, which pops up when the population size is very large compared to the sample size. In this case, it is possible to ignore the 'no replacement' problem and approximate a hypergeometric distribution by the binomial distribution.

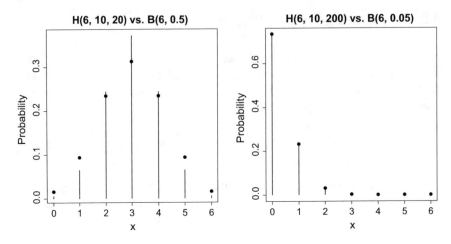

Fig. 3.8 Probability functions of the hypergeometric (*lines*) versus binomial distribution (*dots*). ⊙ BCS_Binhyper

Example 3.6 In Example 3.5, if there are 500 marbles inside the box including 10 red marbles, what is the probability of drawing two red marbles out of 6 draws without replacement?

$$X \sim H(6,\ 10,\ 500),$$

$$P(X = 2) = \frac{\binom{10}{2}\binom{500-10}{6-2}}{\binom{500}{6}} = 0.00507$$

Or it can be approximated by using a binomial distribution with $p = M/N = 0.02$

$$X \sim B(6,\ 0.02)$$

$$P(X = x) = \binom{6}{x}0.02^x(1 - 0.02)^{6-x}$$

$$P(X = 2) \approx 0.00553,$$

which somewhat overestimates the actual value.

The distributions are compared in Fig. 3.8, where the points denote the binomial probability and the vertical lines show the hypergeometric probability. The expectation and variance of the rv X are $E\,X = nM/N$ and $\text{Var}\,X = nM(N - M)(N - n)/\{N^2(N - 1)\}$, respectively.

3.6 Poisson Distribution

Another important distribution is the Poisson distribution, discovered and published by Simeon Denis Poisson in 1837, see Poisson (1837). Later in 1898 Ladislaus von Bortkiewicz made a practical application and showed, that events with low frequency in a large population follow a Poisson distribution even when the probabilities of the events vary, see von Bortkewitsch (1898), Härdle and Vogt (2014).

The Poisson distribution is closely related to the binomial distribution. The example of tossing a die has already been introduced for the binomial distribution, but if we simultaneously toss 100 or 1000 unusual dice with 100 sides each, how can we calculate the probability of a certain number of dice showing the same face? First, we derive the Poisson distribution by taking an approximate limit of the binomial distribution.

Suppose that X is a $B(n, p)$ variable, then the probability mass function is

$$P(X = x) = \frac{n!}{x!(n-x)!} p^x (1-p)^{n-x}, \qquad x = 0, 1, 2, \ldots, n.$$

Let $\lambda = np$, then

$$P(X = x) = \frac{n!}{x!(n-x)!} \left(\frac{\lambda}{n}\right)^x \left(1 - \frac{\lambda}{n}\right)^{n-x} \tag{3.6}$$

$$= \frac{n!}{x!(n-x)!\, n^x} \frac{\lambda^x}{} \frac{(1 - \frac{\lambda}{n})^n}{(1 - \frac{\lambda}{n})^x}$$

$$= \frac{n!}{n^x(n-x)!\, x!} \frac{\lambda^x}{} \frac{(1 - \frac{\lambda}{n})^n}{(1 - \frac{\lambda}{n})^x}.$$

If n is large and p is small, then

$$\frac{n!}{n^x(n-x)!} = \frac{n(n-1)\cdots(n-x+1)}{n^x} \approx 1$$

$$(1 - \lambda/n)^n \approx \exp(-\lambda)$$

$$(1 - \lambda/n)^x \approx 1.$$

Hence (3.6) becomes

$$P(X = x) \approx \exp(-\lambda) \cdot \frac{\lambda^x}{x!}.$$

Eventually, the limiting distribution of the binomial is the Poisson distribution. The comparison is plotted in Fig. 3.9, where the points denote the binomial probability and the vertical lines show the Poisson probability. All the trials are independent and the probability of 'success' is low and equal from trial to trial, while at the same

Fig. 3.9 Probabilities of the Poisson (*lines*) and the Bernoulli distribution (*dots*). ⊡ BCS_Binpois

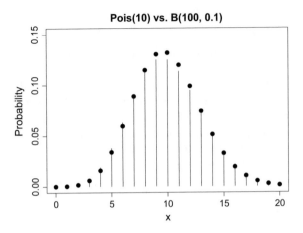

time the total number of trials should be very large. As a rule of thumb, if $p \leq 0.1$, $n \geq 50$ and $np \leq 5$, the approximation is sufficiently close.

Definition 3.11 An rv X follows a *Poisson distribution* with parameter λ, denoted as Pois(λ), if

$$P(X = x) = \exp(-\lambda) \cdot \frac{\lambda^x}{x!}, \qquad x = 0, 1, 2, \ldots \text{ and } \lambda > 0. \tag{3.7}$$

The expectation and the variance of a Poisson rv X is $\mathsf{E}\, X = \lambda$ and $\mathsf{Var}\, X = \lambda$. Note that this does not hold for the binomial distribution.

Example 3.7 If a typist makes on average one typographical error per page, what is the probability of exactly two errors in a one-page text?

Assume that the number of typographical errors per page is an rv $X \sim B(n, p)$. Here n is the number of words per page and p the probability of a word's containing a typographical error. It is plausible to assume that n is large and p small, which is sufficient for X to follow the Poisson distribution with $\lambda = \mathsf{E}\, X = 1$. Using (3.7), the probability of two typographical errors on the same page $P(X = 2)$ is

$$P(X = 2) = \exp(-1) \cdot \frac{1^2}{2!} = 0.184\,.$$

```
> dpois(x = 2, lambda = 1)    # pdf
[1] 0.1839397
```

The parameter λ is also called the intensity, which is motivated by the fact that λ describes the expected number of events within a given interval.

Example 3.8 The Prussian horsekick fatality dataset from Ladislaus von Bortkiewicz (Quine and Seneta 1987) gives us the number of soldiers killed by horsekick in 10

24   **Zweites Kapitel. § 12.**

	75	76	77	78	79	80	81	82	83	84	85	86	87	88	89	90	91	92	93	94
G	—	2	2	1	—	—	1	1	—	3	—	2	1	—	—	1	—	1	—	1
I	—	—	—	2	—	3	—	2	—	—	—	1	1	1	—	2	—	3	1	—
II	—	—	—	2	—	2	—	—	1	1	—	—	2	1	1	—	—	2	—	—
III	—	—	—	1	1	1	2	—	2	—	—	—	1	—	1	2	1	—	—	—
IV	—	1	—	1	1	1	1	—	—	—	1	—	—	—	—	1	1	—	—	—
V	—	—	—	—	2	1	—	—	1	—	—	1	—	1	1	1	1	1	1	—
VI	—	—	1	—	2	—	—	1	2	—	1	1	3	1	1	1	—	3	—	—
VII	1	—	1	—	—	—	1	—	1	1	—	—	2	—	—	2	1	—	2	—
VIII	1	—	—	—	1	—	—	1	—	—	—	—	1	—	—	1	1	1	—	1
IX	—	—	—	—	—	2	1	1	1	—	2	1	1	—	1	2	—	1	—	—
X	—	—	1	1	—	1	—	2	—	2	—	—	—	—	2	1	3	—	1	1
XI	—	—	—	—	2	4	—	1	3	—	1	1	1	1	2	1	3	1	3	1
XIV	1	1	2	1	1	3	—	4	—	1	—	3	2	1	—	2	1	1	—	—
XV	—	1	—	—	—	—	—	1	—	1	1	—	—	—	2	2	—	—	—	—

Fig. 3.10 The original Prussian horsekick dataset

similar corps of the Prussian military over a period of 20 years. Observations on accidents over a total of 200 corps-years are available, as well as $\lambda = 0.61$ estimated from the 200 observations. What is the probability that there was exactly one fatal horsekick over 20 years in a corps? Here we will use both the binomial distribution and the Poisson distribution to calculate the probability, in order to compare the results and see their approximate equality. Since $n = 200$ and $\lambda = 0.61$, $p = \lambda/n = 0.00305$ and the probability can be calculated by (Fig. 3.10).

$$X \sim \text{Pois}(0.61)$$
$$P(X = 1) = \exp(-0.61) \cdot \frac{0.61^1}{1!} = 0.33144,$$
$$X \sim B(200, 0.00305),$$
$$P(X = 1) = \binom{200}{1} 0.00305^1 (1 - 0.00305)^{200-1} = 0.33215.$$

```
> n        = 200
> lambda = 0.61
> p        = lambda / n
> dbinom(x = 1, size = n, prob = p)  # binomial pdf
[1] 0.3321483
> dpois(x = 1, lambda = lambda)       # Poisson pdf
[1] 0.331444
```

3.6.1 Summation of Poisson Distributed Random Variables

Let $X_i \sim \text{Pois}(\lambda_i)$ be independent and Poisson distributed rvs with parameters $\lambda_1, \lambda_2, \ldots, \lambda_n$. Then their sum is an rv also following the Poisson distribution with $\lambda = \lambda_1 + \lambda_2 + \ldots + \lambda_n$:

$$X_i \sim \text{Pois}(\lambda_i), \ i = 1, \ldots, n \text{ implies } \sum_{i=1}^{n} X_i \sim \text{Pois}(\lambda_1 + \lambda_2 + \ldots + \lambda_n).$$

This feature of the Poisson distribution is very useful, since it allows us to combine different Poisson experiments by summing the rates. Furthermore, for two Poisson rvs X and Y, the conditional distribution (see Sect. 6.1) of Y is binomial with the probability parameter $\lambda_Y/(\lambda_X + \lambda_Y)$, see Bolger and Harkness (1965) for more details (Fig. 3.11).

$$\text{If } X \sim \text{Pois}(\lambda_X), \ Y \sim \text{Pois}(\lambda_Y), \text{ then } (X + Y) \sim \text{Pois}(\lambda_X + \lambda_Y)$$

$$\text{and } Y|(X + Y) \sim B\{x + y, \lambda_Y/(\lambda_X + \lambda_Y)\}.$$

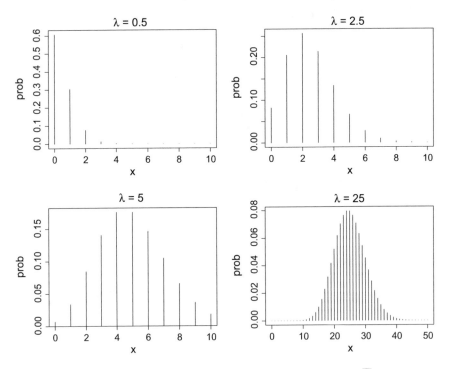

Fig. 3.11 Probability mass functions of the Poisson distribution for different λ. ⌕ BCS_Poispdf

The Poisson distribution belongs to the exponential family of distributions. An rv X follows a distribution belonging to the exponential family if its probability mass function with a single parameter θ has the form

$$P(X = x) = h(x)g(\theta)\exp\{\eta(\theta)t(x)\}.$$

For the Poisson distribution $h(x) = \frac{1}{x!}$, $g(\theta) = \lambda^x$, $\eta(\theta) = -\lambda$ and $t(x) = 1$. Other popular distributions, such as the normal, exponential, gamma, χ^2 and Bernoulli, belong to the exponential family and are discussed in Chap. 4, except for the last. This condition can be extended to multidimensional problems. Furthermore, if we standardise a Poisson rv X, the limiting distribution of this standardised variable follows a standard normal distribution.

$$\frac{X - \lambda}{\sqrt{\lambda}} \xrightarrow{\mathcal{L}} N(0, 1), \quad \text{as } \lambda \to \infty.$$

Chapter 4
Univariate Distributions

Everybody believes in the exponential law of errors [i.e., the
normal distribution]: the experimenters, because they think it
can be proved by mathematics; and the mathematicians,
because they believe it has been established by observation.

— Poincare Henri,"Calcul Des Probabilités."

In this chapter, the theory of discrete random variables from Chap. 3 is extended to continuous random variables. At first, we give an introduction to the basic definitions and properties of continuous distributions in general. Then we elaborate on the normal distribution and its key role in statistics. Finally, we exposit in detail several other key distributions, such as the exponential and χ^2 distributions.

4.1 Continuous Distributions

Continuous random variables (see Definition 3.7) can take on an uncountably infinite number of possible values, unlike discrete random variables, which take on either a finite or a countably infinite set of values. These random variables are characterised by a distribution function and a density function.

Definition 4.1 Let X be a continuous random variable (rv). The mapping $F_X : \mathbb{R} \to [0, 1]$, defined by $F_X(x) = P(X \leq x)$, is called the *cumulative distribution function* (cdf) of the rv X.

Definition 4.2 A mapping $f_X : \mathbb{R} \to \mathbb{R}_+$ is called the *probability density function* (pdf) of an rv X if $f_X(x) = \frac{\partial F_X(x)}{\partial x}$ exists for all x and $\int_{-\infty}^{\infty} f_X(x)\, dx$ exists and takes on the value one.

The cdf of an rv X can now be written as

$$F_X(x) = \int_{-\infty}^{x} f_X(u)\, du.$$

© Springer International Publishing AG 2017
W.K. Härdle et al., *Basic Elements of Computational Statistics*,
Statistics and Computing, DOI 10.1007/978-3-319-55336-8_4

$F_X(b) = \int_{-\infty}^{b} f_X(x)\,dx$ is the probability that the rv X is less than a given value b, and $F_X(b) - F_X(a) = \int_{a}^{b} f_X(x)\,dx$ is the probability that it lies in (a, b).

4.1.1   Properties of Continuous Distributions

As in Chap. 3, the elementary properties of the probability distribution of a continuous rv can be described by an expectation and a variance. The only difference from discrete rvs is in the use of integrals, rather than sums, over the possible values of the rv.

Definition 4.3 Let X be a continuous rv with a density function $f_X(x)$. Then the *expectation* of X is defined as

$$\mathsf{E}\,X = \int_{-\infty}^{\infty} x f_X(x)\,dx. \tag{4.1}$$

The expectation exists if (4.1) is absolutely convergent. It describes the location (or centre of gravity) of the distribution.

Definition 4.4 If the expectation of X exists, the *variance* is defined as

$$\mathsf{Var}\,X = \int_{-\infty}^{\infty} (x - \mathsf{E}\,X)^2 f_X(x)\,dx, \tag{4.2}$$

and the *standard deviation* is $\sigma_X = \sqrt{\mathsf{Var}\,X}$.

The variance describes the variability of the variable and exists if the integral in (4.2) is absolute convergent.

Other useful characteristics of a distribution are its skewness and excess kurtosis. The skewness of a probability distribution is defined as the extent to which it deviates from symmetry. One says that a distribution has negative skewness if the left tail is longer than the right tail of the distribution, so that there are more values on the right side of the mean, and vice versa for positive skewness.

Definition 4.5 The *skewness* of an rv X is defined as

$$S(X) = \mathsf{E}\left\{(X - \mathsf{E}\,X)^3\right\}/\sigma_X^3.$$

The kurtosis is a measure of the peakedness of a probability distribution. The excess kurtosis is used to compare the kurtosis of a pdf with the kurtosis of the normal distribution, which equals 3. Distributions with negative or positive excess kurtosis are called *platykurtic* distributions and *leptokurtic* distributions, respectively.

Definition 4.6 The *excess kurtosis* of an rv X is defined as

$$K(X) = \frac{\mathsf{E}\left\{(X - \mathsf{E}\,X)^4\right\}}{\sigma_X^4} - 3.$$

A complete and unique characterization of the distribution of an rv X is given by its *characteristic function* (cf).

Definition 4.7 For an rv $X \in \mathbb{R}$ with a pdf $f(x)$, the cf is defined as

$$\phi_X(t) = \mathsf{E}\left\{\exp(itX)\right\} = \int_{-\infty}^{\infty} \exp(itx) f_X(x) dx.$$

Unlike discrete distributions, where the characteristic function is also the moment-generating function, the moment-generating function for continuous distributions is defined as the characteristic function evaluated at $-it$. The argument of the distribution is t, which might live in real or complex space.

Definition 4.8 For $X \in \mathbb{R}$ with a pdf $f(x)$, the *moment-generating function* is defined as

$$M_X(t) = \phi_X(-it) = \mathsf{E}\left\{\exp(tX)\right\} = \int_{-\infty}^{\infty} \exp(tx) f_X(x) dx.$$

If the cf is absolutely integrable, then $f_X(t)$ is absolutely continuous and the rv X's pdf is also given by

$$f_X(t) = \frac{1}{2\pi} \int_{-\infty}^{\infty} \exp(-itx) \phi_X(x) dx.$$

Definition 4.9 For any univariate distribution F, and for $0 < p < 1$, the quantity

$$F^{-1}(p) = \inf\{x : F(x) \geq p\}$$

is called the theoretical *pth quantile* or fractile of F, usually denoted as ξ_p and the F^{-1} is called the *quantile function*.

In particular $\xi_{1/2}$ is called the *theoretical median* of F. For the quantile function holds, that it is nondecreasing and left-continuous and satisfies the following inequalities:

i $F^{-1}\{F(x)\} \leq x, \quad -\infty < x < \infty,$
ii $F\{F^{-1}(t)\} \geq t, \quad 0 < t < 1,$
iii $F(x) \geq t$ if and only if $x \geq F^{-1}(t).$

4.2   Uniform Distribution

Continuous uniform distribution or rectangular distribution is a family of symmetric probability distributions such that for each member of the family, all intervals of the same length on the distribution's support are equally probable. The support is defined by the two parameters, a and b, which are its minimum and maximum values. The distribution is often abbreviated $U(a, b)$.

Definition 4.10 An rv U with the pdf

$$f(x, a, b) = \begin{cases} \frac{1}{b-a} & \text{for } a \leq x \leq b, \\ 0 & \text{else.} \end{cases}$$

is said to be uniformly distributed between a and b, and is written as $U \sim U(a, b)$.

An arbitrary rv $X \sim F$ can be converted to a uniform distribution via the probability integral transform:

Definition 4.11 (*Probability Integral Transform*) Suppose $U \sim U(0, 1)$. Then the returned rv $X = F^{-1}(U)$ has the cdf F, i.e. $X \sim F$, and $F(X) \sim U(0, 1)$.

This method can be extended to the discrete case and works even in case of discontinuities in $F(x)$. It is also very often used in the simulation from various distributions, see Chap. 9.

Distribution function and properties of the uniform distribution

The cdf of $X \sim U(a, b)$ is

$$F(x, a, b) = \begin{cases} 0 & \text{for } x < a, \\ \frac{x-a}{b-a} & \text{for } a \leq x \leq b, \\ 1 & \text{for } x > b. \end{cases}$$

The expectation, variance, skewness and excess kurtosis coefficients are

$$\mathsf{E}\,X = \frac{a+b}{2}, \quad \mathsf{Var}\,X = \frac{(b-a)^2}{12}, \quad S(X) = 0, \quad K(X) = -\frac{6}{5}. \tag{4.3}$$

The cf if given through

$$\phi_X(t, a, b) = \frac{e^{itb} - e^{ita}}{it(b-a)}.$$

In order to work with this distribution in R, there is a list of standard implemented functions:

```
dunif(x, min, max), punif(q, min, max),
qunif(p, min, max), runif(n, min, max),
```

which are for the pdf, the cdf, the quantile function and for generating random uniformly distributed samples, respectively. Function `dunif` also contains argument `log` which allows for computation of the log density, useful in the likelihood estimation.

4.3 Normal Distribution

The normal distribution is considered the most prominent distribution in statistics. It is a continuous probability distribution that has a bell-shaped probability density function, also known as the Gaussian function. The normal distribution arises from the central limit theorem, which states, under weak conditions, that the sum of a large number of rvs drawn from the same distribution is distributed approximately normally, irrespective of the form of the original distribution. In addition, the normal distribution can be manipulated analytically, enabling one to derive a large number of results in explicit form. Due to these two aspects, the normal distribution is used extensively in theory and practice.

Definition 4.12 An rv X with the Gaussian pdf

$$\varphi(x, \mu, \sigma^2) = (2\pi\sigma^2)^{-1/2} \exp\left\{-(x - \mu)^2/(2\sigma^2)\right\}$$

is said to be *normally* distributed with $\mathsf{E}\,X = \mu$ and $\mathsf{Var}\,X = \sigma^2$ and is written as $X \sim \mathrm{N}(\mu, \sigma^2)$. If $\mu = 0$ and $\sigma^2 = 1$, then φ is called the *standard normal distribution*, and we abbreviate $\varphi(x, 0, 1)$ as $\varphi(x)$.

An arbitrary normal rv X can be converted to a standard normal distribution, or standardised, by a transformation. The standard normal rv Z is defined as $Z = (X - \mu)/\sigma$ (Fig. 4.1).

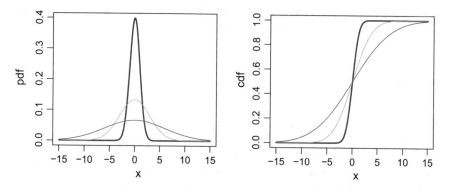

Fig. 4.1 pdf (*left*) and cdf (*right*) of the normal distribution (for $\mu = 0$ and $\sigma^2 = 1, \sigma^2 = 3, \sigma^2 = 6$, respectively). **Q** BCS_NormPdfCdf

Distribution function

The cdf of $X \sim \mathrm{N}(\mu, \sigma^2)$ is

$$\Phi(x, \mu, \sigma^2) = (2\pi\sigma^2)^{-1/2} \int_{-\infty}^{x} \exp\left\{-(u-\mu)^2/(2\sigma^2)\right\} du.$$

Properties of the normal distribution

The cf of the normal distribution is

$$\phi_X(t, \mu, \sigma^2) = \exp\left(i\mu t - \sigma^2 t^2/2\right).$$

The moment-generating function of $X \sim \mathrm{N}(\mu, \sigma^2)$ is

$$M_X(t) = M(t, \mu, \sigma^2) = \exp\left(\mu t + \sigma^2 t^2/2\right).$$

Another useful property of the family of normal distributions is that it is closed under linear transformations. Thus a linear combination of two independent normal rvs, $X_1 \sim \mathrm{N}(\mu_1, \sigma_1^2)$ and $X_2 \sim \mathrm{N}(\mu_2, \sigma_2^2)$, is also normally distributed:

$$aX_1 + bX_2 + c \sim \mathrm{N}(a\mu_1 + b\mu_2 + c, a^2\sigma_1^2 + b^2\sigma_2^2).$$

This property of the normal distribution is actually the direct consequence of a far more general property of the family of distributions called *stable distributions*, see Sect. 4.5.2, as shown in Härdle and Simar (2015).

In order to work with this distribution in R, there is a list of standard implemented functions: dnorm(x, mean, sd), for the pdf (if argument log = TRUE then log density); pnorm(q, mean, sd), for the cdf; qnorm(p, mean, sd), for the quantile function; and rnorm(n, mean, sd) for generating random normally distributed samples. Their parameters are x, a vector of quantiles, p, a vector of probabilities, and n, the number of observations. Additional parameters are mean and sd for the vectors of means and standard deviation, which, if not specified, are set to the standard normal values by default.

4.4   Distributions Related to the Normal Distribution

Central role of the normal distribution in statistics becomes evident when we look at other important distributions constructed from the normal one.

While the normal distribution is frequently applied to describe the underlying distribution of a statistical experiment, asymptotic test statistics (see Sect. 5.2.2) are often based on a transformation of a (non-) normal rv. To get a better understanding of these tests, it will be helpful to study the χ^2, t- and F-distributions, and their relations with the normal one. Skew or leptokurtic distributions, such as the exponential, stable and Cauchy distributions, are commonly required for modelling extreme events or an rv defined on positive support, and therefore will be discussed subsequently.

4.4.1 χ^2 Distribution

In statistics, the χ^2 distribution describes the sum of the squares of independent standard normal rvs.

Definition 4.13 If $Z_i \sim N(0, 1), i = 1, ..., n$ are independent, then the rv X given by

$$X = \sum_{i=1}^{n} Z_i^2 \sim \chi_n^2$$

is χ^2 distributed with n degrees of freedom.

This distribution is of particular interest since it describes the distribution of a sample variance (see Sect. 5.1.6) and is used further in tests (see Sect. 5.2.2).

Density function

The pdf of the χ^2 distribution is

$$f(z, n) = \frac{2^{-n/2} z^{n/2-1} \exp(-z/2)}{\Gamma(n/2)},$$

where $\Gamma(k)$ is the *Gamma function* $\Gamma(z) = \int_0^\infty t^{z-1} \exp(-t) dt$.

Distribution function

The cdf of the χ^2 distribution is

$$F(z, n) = \frac{\Gamma_z(n/2, z/2)}{\Gamma(n/2)},$$

where Γ_z is the incomplete Gamma function: $\Gamma_z(\alpha) = \int_0^z t^{\alpha-1} \exp(-t) dt$.
   In order to work with this distribution in R, there is a list of standard implemented functions:

```
dchisq(x, df), pchisq(q, df), qchisq(p, df), rchisq(n, df),
```

which are for the pdf, the cdf, the quantile function and for generating random χ^2-distributed samples, respectively. Same as for other distributions, if `log = TRUE` in `dchisq` function, then log density is computed, which is useful for maximum likelihood estimation. Similar to the functions for the t (see Sect. 4.4.2) and F (see Sect. 4.4.3) distributions, all the functions also have the parameter `ncp` which is the non-negative parameter of non-centrality, where this rv is constructed from Gaussian rvs with non-zero expectations.

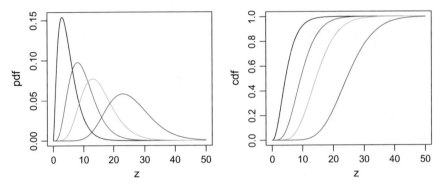

Fig. 4.2 pdf (*left*) and cdf (*right*) of χ^2 distribution (degrees of freedom $n = 5$, $n = 10$, $n = 15$, $n = 25$, respectively). $\textbf{Q}$ BCS_ChiPdfCdf

Fig. 4.3 Pdf of χ^2 distribution ($n = 1$ and $n = 2$). $\textbf{Q}$ BCS_ChiPdf

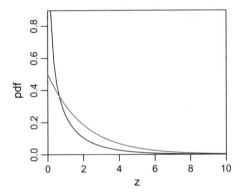

Figure 4.2 illustrates the different shapes of the χ^2 distribution's cdf and pdf, for different degrees of freedom n. In general, the χ^2 pdf is bell-shaped and shifts to the right-hand side for greater numbers of degrees of freedom, becoming more symmetric.

There are two special cases, namely $n = 1$ and $n = 2$. In the first case, the vertical axis is an asymptote and the distribution is not defined at 0. In the second case, the curve steadily decreases from the value 0.5 (Fig. 4.3).

Properties of the χ^2 distribution

A distinctive feature of χ^2 is that it is positive, due to the fact that it represents a sum of squared values.

The expectation, variance, skewness and excess kurtosis coefficients are

$$\text{E}\,X = n, \quad \text{Var}\,X = 2n, \quad S(X) = 2\sqrt{\frac{2}{n}}, \quad K(X) = \frac{12}{n}.$$

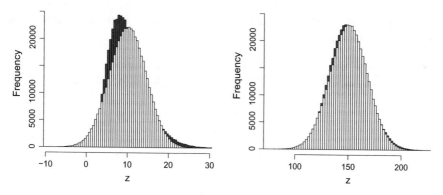

Fig. 4.4 Asymptotic normality of χ^2 distribution (*left panel n* = 10; *right panel n* = 150).
⌕ BCS_ChiNormApprox

A χ^2-distributed variable X reaches its maximum value when $z = n - 2$, given that the number of degrees of freedom is $n \geq 2$. Interestingly, it follows that if X is χ^2-distributed with $n \to \infty$ degrees of freedom, then (in an asymptotic sense)

(i) $X \xrightarrow{\mathcal{L}} N(n, 2n)$

(ii) $\sqrt{2X} \xrightarrow{\mathcal{L}} N(\sqrt{2n - 1}, 1)$.

In order to check the asymptotic property in (ii), we have generated samples of such two types:

```
> # samples for n = 1
> x1      = rchisq(n = 500000,
+    df = 1)                    # chi-square distr. with 1 df
> norm1 = rnorm(n = 500000,     # normal distr.
+    mean = 1,                  # with expectation = 1
+    sd  = sqrt(2 * 1))         # and variance = 2
> # samples for n = 2
> x2      = rchisq(n = 500000,
+    df = 2)                    # chi-square distr. with 2 df
> norm2 = rnorm(n = 500000,     # normal distr.
+    mean = 2,                  # with expectation = 2
+    sd  = sqrt(2 * 2))         # and variance = 4
```

One can observe in Fig. 4.4 that the χ^2 distribution (coloured in blue) approaches the standard normal distribution for large numbers of degrees of freedom.

4.4.2   Student's t-distribution

A combination of the normal and χ^2 distributions is represented by the *t*-distribution. It gained importance because it is widely used in statistical tests, particularly in Student's *t*-test for estimating the statistical significance of the difference between two sample means. It is also used to construct confidence intervals for population means and linear regression analysis.

Definition 4.14 Let $X \sim N(0, 1)$ and $Y \sim \chi_n^2$ be independent rvs, then a t-distributed rv Z with $n - 1$ degrees of freedom can be formalised as

$$Z = \frac{X}{\sqrt{Y/n}} \sim t_{n-1}.$$

The noncentral t-distribution is a generalised version of Student's distribution:

$$Z = \frac{X + \mu}{\sqrt{Y/n}},$$

where μ is a non-centrality parameter and $X \sim N(0, 1)$ and $Y \sim \chi_n^2$ are independent.

Density function

The pdf of the t-distribution is

$$f(z, n) = \frac{\Gamma\{(n + 1)/2\}}{\sqrt{\pi n}\, \Gamma(n/2) \left(1 + z^2/n\right)^{(n+1)/2}}.$$

Distribution function

The cdf of the t-distribution is

$$F(z) = \int_{-\infty}^{z} f(t, n)dt = \frac{B(z; n/2; n/2)}{B(n/2; n/2)},$$

where $B(n/2; n/2)$ is the *Beta* function: $B(x, y) = \int_0^1 t^{x-1}(1 - t)^{y-1}dt$ and $B(z; n/2; n/2)$ is the incomplete *Beta* function $B(z; a, b) = \int_0^z t^{a-1}(1 - t)^{b-1}dt$.
    Similar to other distributions, the R functions for t-distribution are

```
dt(x, df), pt(q, df), qt(p, df), rt(n, df),
```

for computing the pdf, cdf, quantile function and generating random numbers. Same as for other distributions, if log = TRUE in dt function, then log density is computed, which is useful for maximum likelihood estimation. Also similar to the functions for the χ^2 and F (see Sect. 4.4.3) distributions, all the above-mentioned functions have the non-centrality parameter ncp.
    Figure 4.5 shows the standard normal distribution (black bold line) and several different t-distributions with different degrees of freedom.

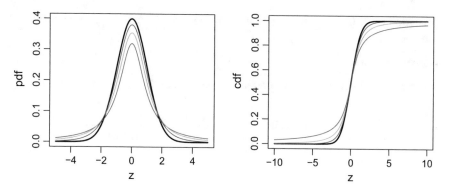

Fig. 4.5 Density function of Student's t-distribution and correspondent cumulative distribution functions ($n = 1, n = 2, n = 5$, *bold line*- N(0, 1)). **Q** BCS_tPdfCdf

Properties of Student's t-distribution

For $n > 2$ degrees of freedom, the expectation and variance of Student's t-distribution are

$$E\,Z = 0,\ \ \text{Var}\,Z = \frac{n}{n-2},$$

otherwise they do not exist.

The skewness and excess kurtosis are

$$S(Z) = 0, \quad K(Z) = \frac{3(n-2)}{n-4}, \quad \text{for } n > 4.$$

The quantiles of a t-distributed rv Z are denoted by t_p, and, due to symmetry, $t_p = -t_{1-p}$. Thus, when stating the hypothesis for a two-sided test, the critical values are used with given significance levels α as follows:

$$P(|x| > |t_{1-\alpha/2}|) = \alpha.$$

4.4.3   F-distribution

Definition 4.15   The rv Z has the *Fisher–Snedecor* (F-distribution) distribution with n and m degrees of freedom if

$$Z = \frac{\chi^2(n)/n}{\chi^2(m)/m} \sim F_{n,m},$$

where $\chi^2(n) \sim \chi^2_n$ and $\chi^2(m) \sim \chi^2_m$ are independent rvs.

This kind of distribution is directly used in analysis of variance problems (F-test), see Sect. 5.2.2.

Density function

The pdf of an F-distributed rv contains the degrees of freedom n and m:

$$f(z, n, m) = (n/m)^{n/2} \frac{\Gamma\{(n+m)/2\} z^{(n/2-1)}}{\Gamma(n/2) \Gamma(m/2) (1+nz/m)^{(n+m)/2}}.$$

Distribution function

The cdf is

$$F(z) = 2n^{(n-2)/2} (x/m)^{n/2} \frac{F_h\{(n+m)/2, n/2; 1+n/2; -nz/m\}}{B(n/2, m/2)} \quad \text{for } z \geq 0,$$

where F_h is the hypergeometric function.

The procedures in R dedicated to this distribution require the parameters n and m as well:

```
df(x, df1, df2), pf(q, df1, df2),
qf(p, df1, df2), rf(n, df1, df2),
```

for computing the pdf, cdf, quantile function and generating random numbers. Here parameters df1 and df2 are the two degrees of freedom parameters. Same as for other distributions, if log = TRUE in df function, then log density is computed, which is useful for maximum likelihood estimation. Also similar to the functions for the χ^2 and t-distribution, all the above-mentioned functions have the non-centrality parameter ncp.

Distribution parameters

The expectation and variance of the F-distribution are defined if $m > 2$:

$$\mathrm{E}\,Z = \frac{m}{m-2}, \quad \mathrm{Var}\,Z = \frac{2m^2(n+m-2)}{n(m-2)^2(m-4)}.$$

And the skewness coefficient expression is

$$S(Z) = \frac{(2n+m-2)\sqrt{8(m-4)}}{(m-6)\sqrt{n(n+m-2)}} \quad \text{for } m > 6.$$

Looking at Fig. 4.6, one can distinguish three characteristic shapes of the pdf curve, depending on the parameters n and m:

- for $n = 1$, the curve monotonically decreases for all values of m with the vertical axis as an asymptote;

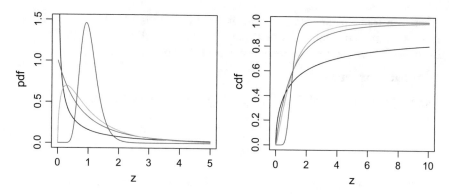

Fig. 4.6 Density and cumulative distribution of the F-distribution ($n = 1, m = 1, n = 2, m = 6,$ $n = 3, m = 10$ and $n = 50, m = 50$, respectively). $\mathbf{\mathsf{Q}}$ BCS_FPdfCdf

- for $n = 2$, the curve again decreases for all m, but intersects the vertical axis at the point 1;
- for $n \geq 3$, the curve has an asymmetrical bell shape for all m, gradually shifting to the right-hand side for larger numbers of degrees of freedom.

4.5   Other Univariate Distributions

4.5.1   Exponential Distribution

Example 4.1 Let us assume that over the time interval $[0, T]$, the online service of a food delivery company receives x orders. At some point, the managers of this business became curious as to the probabilities of the amounts of orders over time.

In general, the number of orders can be described by a Poisson distribution, see Definition 3.7, where λ is the expected number of occurrences during a given time period. If during one hour the online service receives on average $\lambda = 35$ orders, then within any given hour the probability of receiving exactly 30 orders has a probability of $p = 35^{30}e^{-35}/30! = 0.049$.

However, when we need to model the distribution of time intervals between orders, or events, the *exponential distribution* comes in handy.

Density function

The pdf of the exponential distribution is defined as

$$f(z, \lambda) = \begin{cases} \lambda e^{-\lambda z}, & \text{for } z \geq 0, \\ 0, & \text{for } z < 0, \end{cases}$$

where λ is a rate parameter, such that the time interval is $1/\lambda$. The rate parameter gives the expected number of events in a time interval, whereas its reciprocal gives the expected time interval between two events. And one writes $X \sim \mathcal{E}(\lambda)$.

Distribution function

The expression for the cdf looks relatively similar to that of the pdf:

$$F(x, \lambda) = \begin{cases} 1 - e^{-\lambda x}, & \text{for } x \geq 0, \\ 0, & \text{for } x < 0. \end{cases}$$

In general, the greater the λ is, the steeper are the curves of the exponential density and distribution functions (Fig. 4.7).

The main R functions for the exponential distribution are

```
dexp(x, rate), pexp(q, rate),
qexp(p, rate), rexp(n, rate).
```

for computing the pdf, cdf, quantile function and generating random numbers. Same as for other distributions, if `log = TRUE` in `dexp` function, then log density is computed, which is useful for maximum likelihood estimation.

Example 4.2 University beverage vending machines have a lifetime of X, which is exponentially distributed with $\lambda = 0.3$ defective machines per year:

$$f(x, 0.3) = \begin{cases} 0.3e^{-0.3x}, & \text{for } x \geq 0, \\ 0, & \text{for } x < 0. \end{cases}$$

We would like to find the probability that this vending machine will function more than 1.7 years.

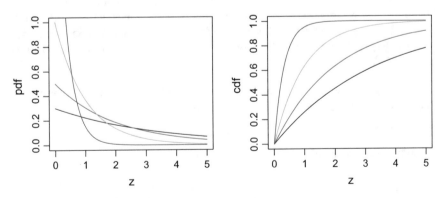

Fig. 4.7 Pdf and cdf of the exponential distribution ($\lambda = 0.3$, $\lambda = 0.5$, $\lambda = 1$ and $\lambda = 3$).
Q BCS_ExpPdfCdf

The reliability after time x is given by

$$P(X > x) = e^{-0.3x}.$$

```
> 1 - pexp(q = 1.7, rate = 0.3)
[1] 0.6004956
```

Thus the probability that a breakdown occurs within 1.7 years is approximately 60%.

Properties of the exponential distribution

The exponential distribution has the following expectation and variance:

$$\mathsf{E}\,X = 1/\lambda, \quad \mathsf{Var}\,X = 1/\lambda^2.$$

The mode (see Definition 5.6) is 0 and the median (see Definition 4.9) is

$$\xi_{1/2} = \frac{\log 2}{\lambda}.$$

The exponential distribution has skewness and excess kurtosis coefficients independent of λ, unlike some of the distributions we have seen so far. They are

$$S(X) = 2, \quad K(X) = 6.$$

Another interesting property of the exponential distribution is that it is memoryless, something which can be said of only two distribution: the exponential and the geometric. It follows that the expected time until the next event is constant and does not depend on the time elapsed since the occurrence of the previous event. We can see this property in a more formal presentation through

$$P\,(X \le t + q | X > t) = P\,(X \le q).$$

The conditional probability of the next event's occurring by time $t + q$ given that the last event was at time t is equal to the unconditional probability of the next event's occurring at time q without any previous information.

4.5.2 Stable Distributions

As mentioned in Sect. 4.3, the stable distributions are a family of distributions which are closed under linear transformations.

Definition 4.16 A distribution function is said to be *stable* if for any two independent rvs Z_1 and Z_2 following this distribution, and any two positive constants a and b, we have

$$aZ_1 + bZ_2 = cZ + d,$$

where c is a positive constant, d is a constant, and Z is an rv with the same distribution as Z_1 and Z_2. Here, c and d depend on a and b.

If this property holds with $d = 0$, then such a distribution is said to be strictly stable.

In order to completely define stable distributions, we need four parameters: the index of stability $\alpha \in (0, 2]$ (determines the thickness of the tails), the skewness parameter $\beta \in [-1, 1]$ (determining the asymmetry), the location parameter $\mu \in \mathbb{R}$ and a scale parameter $\sigma > 0$.

Figure 4.8 shows the effect of the parameters α and β on the shape of the density curves. Greater values of α make the peak less pointed and the tails fatter. For $\beta > 0$, the distribution is skewed to the right (the right tail is fatter) and for $\beta < 0$, it is skewed to the left. When $\beta = 0$, the distribution is symmetric around the peak.

The pdf and cdf of stable distributions

Generally, the pdf and the cdf of stable distributions cannot be written down analytically (except for three special cases). However, stable distributions can be described by their cf ϕ_Z.

A conventional parameterization of the cf of a stable rv Z with parameters α, β, σ_S and μ is given by Samorodnitsky and Taqqu (1994), Weron (2001):

$$\log \phi_Z(t, \alpha, \beta, \sigma_S, \mu) = \begin{cases} -\sigma_S^\alpha |t|^\alpha (1 - i\beta \mathrm{sign}(t) \tan \frac{\pi\alpha}{2}) + i\mu t, & \alpha \neq 1, \\ -\sigma_S |t| (1 + i\beta \mathrm{sign}(t) \frac{2}{\pi} \log |t|) + i\mu t, & \alpha = 1, \end{cases}$$

where $i^2 = -1$. Note that the σ used here is not the usual Gaussian scale σ, but the value $\sigma_S = \sigma/\sqrt{2}$.

In R the stable distributions can be implemented by

```
dstable(z, alpha, beta, gamma, delta, pm)
```

and

```
pstable(z, alpha, beta, gamma, delta, pm),
```

which require the `stabledist` package. We can easily work with a stable distribution of interest that depends on the parameters α, β, σ, μ and the parameter pm, which refers to the parameterization type. The functions `qstable` and `rstable` with the same parameters let us use quantiles and generate samples.

An interesting toolbox is implemented with the command `stableSlider()` (of the `fBasics` package). It provides a good illustration of the pdf and cdf functions of different stable distributions. One can change the parameters to see how the shape of the functions reacts to the changed values, see Fig. 4.9. There exist three special cases of stable distributions that have closed form formulas for their pdf and cdf:

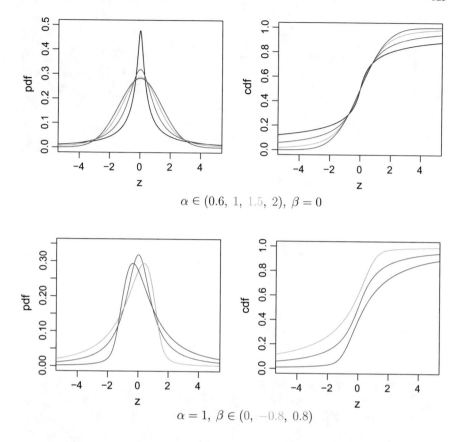

$\alpha \in (0.6,\ 1,\ 1.5,\ 2),\ \beta = 0$

$\alpha = 1,\ \beta \in (0,\ -0.8,\ 0.8)$

Fig. 4.8 Stable distribution functions and their density functions given different combinations of α and β (in all cases $\sigma = 1$ and $\mu = 0$). **Q** BCS_StablePdfCdf

Normal Distribution
$$f(z) = \frac{1}{\sqrt{2\pi\sigma^2}} \exp\left\{-\frac{(z-\mu)^2}{2\sigma^2}\right\},$$

Cauchy Distribution
$$f(z) = \sigma / \left\{\pi(z-\mu)^2 + \pi\sigma^2\right\}, \qquad (4.4)$$

Lévy Distribution
$$f(z) = \sqrt{\frac{c}{2\pi}} \frac{\exp\left\{-\frac{c}{2(z-\mu)}\right\}}{(z-\mu)^{3/2}}.$$

With the help of the following short code we plot the pdf for those special cases. These can be built using the `dstable` function from package `stabledist` with the appropriate parameters α, β, σ and μ (Fig. 4.10).

```
> require(stabledist)
> z        = seq(-6, 6, length = 300)
> s.norm = dstable(z,       # values of the density
+    alpha = 2,             # tail
+    beta  = 0,             # skewness
```

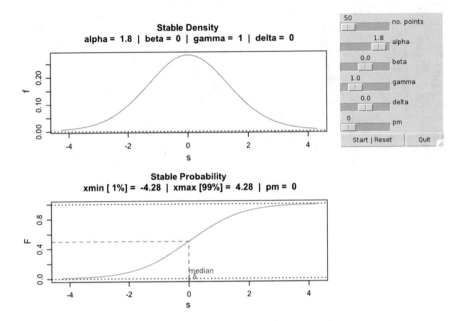

Fig. 4.9 Screenshot of fBasics package's stableSlider() toolbox

```
+    gamma = 1,          # scale
+    delta = 0,          # location
+    pm    = 1),         # type of parametrization
> s.cauchy = dstable(z,  # values of the density
+    alpha = 1,          # tail
+    beta  = 0,          # skewness
+    gamma = 1,          # scale
+    delta = 0,          # location
+    pm    = 0),         # type of parametrization
> s.levy   = dstable(z,  # values of the density
+    alpha = 0.5,        # tail
+    beta  = 0.9999,     # skewness
+    gamma = 1,          # scale
+    delta = 0,          # location
+    pm    = 0),         # type of parametrization
> plot(z, s.norm,        # plot normal
+    col ="red", type ='l', ylim = c(0,0.5))
> lines(z, s.cauchy,     # plot Cauchy
+    col ="green")
> lines(z, s.levy,       # plot Levy
+    col ="blue")
```

In all cases, $\sigma = 1$ and $\mu = 0$. The cdf functions can be plotted analogously using the procedure pstable.

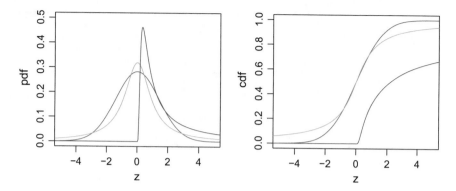

Fig. 4.10 Special cases of stable distributions (Gaussian: $\alpha = 2$, $\beta = 0$; Cauchy: $\alpha = 1, \beta = 0$; Lévy: $\alpha = 0.5$, $\beta = 1$). $\mathbf{Q}$ BCS_StablePdfCdfSpecial

4.5.3 Cauchy Distribution

The Cauchy–Lorentz distribution is a continuous probability distribution which is stable and known for having an undefined mean and an infinite variance. It is important in physics since it is the solution of differential equations describing forced resonance and describes the shape of spectral lines.

Example 4.3 Consider an isotropic source emitting particles to the plane L. The angle θ of each emitted particle is uniformly distributed. Each particle hits the plane at some distance x from the point 0 (Fig. 4.11). By definition, the distance rv X follows Cauchy distribution.

Density function

The pdf of the Cauchy distribution is defined as in (4.4) where $\mu \in \mathbb{R}$ is a location parameter, i.e. it defines the position of the peak of the distribution, and $\sigma > 0$ is a scale parameter specifying one-half the width of the probability density function at one-half its maximum height. For $\mu = 0$ and $\sigma = 1$, the distribution is called a standard Cauchy distribution (Fig. 4.12).

Fig. 4.11 Illustration of the generation of a Cauchy-distributed rv (Particles are emitted with uniformly distributed angles and hit the plane L so that the distance x is Cauchy distributed

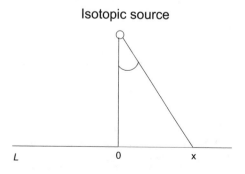

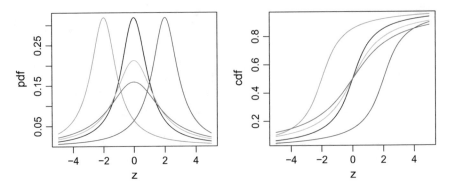

Fig. 4.12 Cauchy distribution functions and corresponding density functions ($\mu = -2$, $\sigma = 1$; $\mu = 0$, $\sigma = 1$; $\mu = 2$, $\sigma = 1$; $\mu = 0$, $\sigma = 1.5$; $\mu = 0$, $\sigma = 2$). **Q** BCS_CauchyPdfCdf

Distribution function

The Cauchy cdf is

$$F(z; \mu, \sigma) = \frac{1}{\pi} \arctan\left(\frac{z - \mu}{\sigma}\right) + \frac{1}{2}.$$

In R, the pdf, cdf, quantile function and generating random numbers from Cauchy distribution can be done using the commands

```
dcauchy(z, mu, sigma), pcauchy(p, mu, sigma),
qcauchy(p, mu, sigma), rcauchy(q, mu, sigma),
```

or by using the

```
dstable, pstable, qstable, rstable
```

functions with parameters $\alpha = 1$ and $\beta = 0$.

Properties of the Cauchy distribution

As mentioned before, the Cauchy distribution has a non-finite expectation. As a consequence, its variance, skewness, kurtosis and other higher order moments do not exist. Its mode (see Definition 5.6) and the median (see Definition 4.9) are both defined and equal to μ.

Cauchy rvs can be simply thought of as a ratio of two $N(0, 1)$ rvs: if $X \sim N(0, 1)$ and $Y \sim N(0, 1)$ are independent rvs, then X/Y follows the standard Cauchy distribution.

The Cauchy distribution is an infinitely divisible distribution, which means that for any positive n, there exist n independent and identically distributed rvs $X_{n1}, \ldots, X_{nn}$ whose sum has the Cauchy distribution.

It is worth mentioning that the standard Cauchy distribution coincides with Student's t-distribution with one degree of freedom.

Chapter 5
Univariate Statistical Analysis

It is a capital mistake to theorise before one has data. Insensibly one begins to twist facts to suit theories, instead of theories to suit facts.

— Sir Arthur Conan Doyle

This chapter presents basic statistical methods used in describing and analysing univariate data in R. It covers the topics of descriptive and inferential statistics of univariate data, which are mostly treated in introductory courses in Statistics.

Among other useful statistical tools, we discuss simple techniques of explorative data analysis, such as the Bar Diagram, Bar Plot, Pie Chart, Histogram, kernel density estimator, the ecdf, and parameters of location and dispersion. We also demonstrate how they are easily implemented in R. Further in this chapter we discuss different test for location, dispersion and distribution.

5.1 Descriptive Statistics

Let us consider an rv X following a distribution F_θ, $X \sim F_\theta$. To obtain a sample of n observations one first constructs n copies of the rv X, i.e. *sample rvs* $X_1, \ldots, X_n \sim F_\theta$ which follow the same distribution F_θ as the original variable X. All $X_1, \ldots, X_n$ are often assumed to be independent. Subsequently, we draw one observation x_i from every sample rv X_i; this results in a random sample $x_1, \ldots, x_n$. Note that these are not rvs but numbers and can be used to estimate the unknown parameters θ (for normal distribution these are μ and σ). Thus, all the functions based on the sample discussed later in this chapter might be random as well as nonrandom. For example, sample mean (see Sect. 5.1.5) $\bar{x} = \frac{1}{n}\sum_{i=1}^{n} x_i$ is non-random and defines the center of the cloud of observations. On the other hand $\bar{X} = \frac{1}{n}\sum_{i=1}^{n} X_i$ is a rv, which can be characterized by a distribution function and has the property, $\mathsf{E}\bar{X} = \mathsf{E}X$.

© Springer International Publishing AG 2017
W.K. Härdle et al., *Basic Elements of Computational Statistics*,
Statistics and Computing, DOI 10.1007/978-3-319-55336-8_5

Having realizations $\{x_i\}$, $i \in 1, \ldots, n$ of a rv X, with $\{a_j\}$, $j \in 1, \ldots, k$, denoting all possible but different realizations of X in the sample, we can define the following two types of frequencies.

Definition 5.1 The *absolute frequency* of a_j, denoted by $n(a_j)$, is the number of occurrences of a_j in the sample $\{x_i\}$. The *relative frequency* of a_j, denoted by $h(a_j)$, is the ratio of the absolute frequencies of a_j and the sample size n: $h(a_j) = n(a_j)/n$. Clearly $\sum_{j=1}^{k} h(a_j) = 1$.

The function `table()` returns all possible observed values of the data along with their absolute frequencies. These can be used further to compute the relative frequencies by dividing by n.

Let us consider the dataset `chickwts`, a data frame with 71 observations of 2 variables, `weight`, a numeric variable for the weight of the chicken, and `feed`, a factor for the type of feed. In order to select only the observed values of `feed`, one considers the field `chickwts$feed`. By using `table(chickwts$feed)`, we get one line, stating the possible chicken feed, i.e. each possible observational value, and the absolute frequency of each type in the line below.

```
> table(chickwts$feed)              # absolute frequencies

   casein horsebean    linseed  meatmeal    soybean sunflower
       12        10         12        11         14        12
> n = length(chickwts$feed); n      # sample size
[1] 71
> table(chickwts$feed) / n          # relative frequency

   casein horsebean    linseed  meatmeal    soybean sunflower
0.1690141 0.1408451 0.1690141 0.1549296 0.1971831 0.1690141
```

5.1.1   Graphical Data Representation

There are several methods of visualising the frequencies graphically. Depending on which type of data is at our disposal, the adequate approach needs to be chosen carefully: Bar Plot, Bar Diagram or Pie Chart for qualitative or discrete variables, and the histogram for continuous (or quasi-continuous) variables.

Since the variable `feed` in dataset `chickwts` has only 6 distinct observed values of `feed`, the frequencies can be conveniently shown using a Bar Plot, a Bar Diagram or a Pie Chart.

Bar diagram

In a Bar Diagram, each observation is plotted using sticks. The y-axis indicates, depending on the specification, the absolute or relative frequencies.

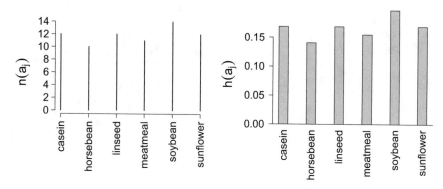

Fig. 5.1 Bar diagram of the absolute frequencies $n(a_j)$ (*left*) and bar plot of the relative frequencies $h(a_j)$ (*right*) of `chickwts$feed`. **Q** BCS_BarGraphs

```
> n = length(chickwts$feed)       # sample size
> plot(table(chickwts$feed))      # absolute frequency
> plot(table(chickwts$feed) / n)  # relative frequency
```

The result of the first `plot` command is shown in the left panel of Fig. 5.1.

Bar plot

Unlike in the Bar Diagram, each observation is plotted using bars in the Bar Plot. If the endpoints of the bars are connected, one obtains a *frequency polygon*. It is in particular useful to illustrate the behaviour (variation) of time ordered data.

```
> n = length(chickwts$feed)          # sample size
> barplot(table(chickwts$feed))      # absolute frequency
> barplot(table(chickwts$feed) / n)  # relative frequency
```

The result of the second `baplot` command is shown in the right panel of Fig. 5.1.

Pie chart

In a Pie Chart, each observation has its own sector with an angle (or a square for a square Pie Chart) proportional to its frequency. The angle can be obtained from $\alpha(a_i) = h(a_i) \cdot 360°$. The disadvantage of this approach is that the human eye cannot precisely distinguish differences between angles (or areas). Instead, it recognises much better differences in lengths, which is the reason why the Bar Plot and Bar Diagram are better tools than the Pie Chart. In Fig. 5.2, each group seems to have the same area in the Pie Chart, though the frequencies differ slightly from each other, as is evident from the Bar Plot in Fig. 5.1.

```
> pie (table(chickwts$feed))
```

Fig. 5.2 Pie chart of the
data chickwts$feed.
Q BCS_pie

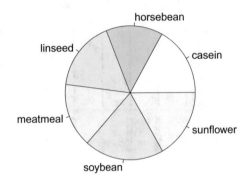

5.1.2 Empirical (Cumulative) Distribution Function

The empirical cumulative distribution function (ecdf) is denoted by $\hat{F}(x)$ and describes the *relative number of observations which are less than or equal to x in the sample*. We write $\hat{F}(x)$ with a hat because it is an estimator of the true cdf $F(x) = P(X \leq x)$ (see Definition 3.4 for discrete rvs and Definition 4.1 for continuous rvs), as for every value of x, $\hat{F}(x)$ converges almost surely to $F(x)$ when n goes to infinity (*Strong Law of Large Numbers*, Serfling 1980), i.e.

$$\hat{F}(x) = n^{-1} \sum_{i=1}^{n} \mathrm{I}(X_i \leq x) \xrightarrow{a.s.} F(x). \tag{5.1}$$

The ecdf is a non-decreasing *step function*, i.e. it is a function which is constant except for jumps at a discrete set of points. The points where the jumps occur are the realisations of the rv and thus illustrate a general property of the cumulative distribution function: *continuous at right* and *existing limit at left*, which holds for all $\hat{F}$. This step function is given by ecdf(), which returns a function of class ecdf, which can be plotted using plot().

Take for example the dataset Formaldehyde containing 6 observations on two variables: carbohydrate (car) and optical density (optden). The ecdf for the subset Formaldehyde$car can be calculated as follows, with the result being shown in Fig. 5.3.

```
> ecdf (Formaldehyde$car)        # ecdf
Empirical CDF
Call: ecdf(Formaldehyde$car)
 x[1:6] =     0.1,     0.3,     0.5,    ...,     0.7,     0.9
> plot(ecdf(Formaldehyde$car))   # plot of ecdf
```

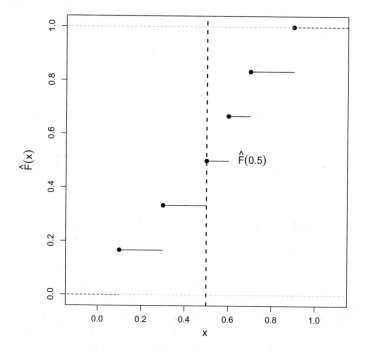

Fig. 5.3 ecdf of `Formaldehyde$car`. **Q** BCS_ecdf

5.1.3 Histogram

The histogram is a common way of visualising the data frequencies of continuous (real valued) variables. In a histogram, bars are erected on distinct *intervals*, which constitute the so-called *classes*. The y-axis represents the relative frequency of the classes, so that the total area of the bars is equal to one. While the width of the bars is given by the chosen intervals, the height of each bar is equal to the *empirical density* of the corresponding interval.

If $\{K_i\}, i = 1, ..., s$ is a set of s disjunct classes, the histogram or empirical density function $\hat{f}(x)$ is defined by

$$\hat{f}(x) = \frac{\text{relative frequency of the class containing } x}{\text{length of the class containing } x} = \frac{h(K_i)}{|K_i|} \text{ for } x \in K_i, \quad (5.2)$$

where $|K_i|$ denotes the length of the class K_i and $h(K_i)$ its relative frequency, which is calculated as the ratio of the number of observations falling into class K_i to the sample size n.

We write $\hat{f}(x)$ with a hat because it is a sample-based estimator of the true density function $f(x)$, which describes the relative likelihood of the underlying variable to take on any given value. $\hat{f}(x)$ is a consistent estimator of $f(x)$, since for every value

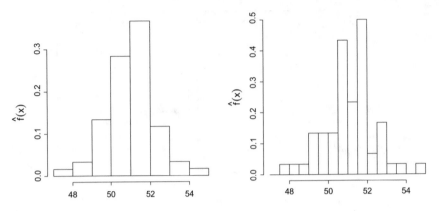

Fig. 5.4 Histograms of nhtemp with the number of classes calculated using default method (*left*) and by manually setting to intervals of length 0.5 (*right*). $\mathbb{Q}$ BCS_hist1, $\mathbb{Q}$ BCS_hist2

of x, $\hat{f}(x)$ converges almost surely to $f(x)$ when n goes to infinity (*Strong Law of Large Numbers*, Serfling 1980).

Now, consider nhtemp, a sample of size $n = 60$ containing the mean annual temperature in degrees Fahrenheit in New Haven, Connecticut, from 1912 to 1971. The histograms in Fig. 5.4 are produced using the function hist(). By default, without specifying the arguments for hist(), R produces a histogram with the absolute frequencies of the classes on the y-axis. Thus, to obtain a histogram according to our definition, one needs to set freq = FALSE. The number of classes s is calculated by default using *Sturges' formula* $s = \lceil \log_2 n + 1 \rceil$. The brackets denote the *ceiling function* used to round up to the next integer (see Sect. 1.4.1) to avoid fractions of classes. Note that this formula performs poorly for $n < 30$. To specify the intervals manually, one can fill the argument breaks with a vector giving the breakpoints between the histogram cells, or simply the desired number of cells. In the following example, breaks = seq(47, 55, 0.5) means that the histogram should range from 47 to 55 with a break every 0.5 step, i.e. $K_1 = [47, 47.5)$, $K_2 = [47.5, 48)$, ...

```
> hist(nhtemp, freq = FALSE)
> hist(nhtemp, freq = FALSE, breaks = seq(47, 55, 0.5))
```

Figure 5.4 displays histograms with different bin sizes. A better reflection of the data is achieved by using more bins. But, as the number of bins increases, the histogram becomes less smooth. Finding the right level of smoothness is an important task in nonparametric estimation, and more information can be found in Härdle et al. (2004).

5.1.4 Kernel Density Estimation

The histogram is a density estimator with a relatively low rate of convergence to the true density. A simple idea to improve the rate of convergence is to use a function that weights the observations in the vicinity of the point where we want to estimate the density, depending on how far away each such observation is from that point. Therefore, the estimated density is defined as

$$\hat{f}_h(x) = \frac{1}{n} \sum_{i=1}^n K_h(x - x_i) = \frac{1}{nh} \sum_{i=1}^n K\left(\frac{x - x_i}{h}\right),$$

where $K\left(\frac{x-x_i}{h}\right)$ is the kernel, which is a symmetric nonnegative real valued integrable function. Furthermore, the kernel should have the following properties:

$$\int_{-\infty}^{\infty} u K(u) du = 0,$$

$$\int_{-\infty}^{\infty} K(u) du = 1.$$

These criteria define a pdf and it is straightforward to use different density functions as a kernel. This is the basic idea of kernel smoothing. The foundations in this area were laid in Rosenblatt (1956) and Parzen (1962). Some examples for different weight functions are given in Fig. 5.5 and Table 5.1.

Deriving a formal expression for the kernel density estimator is fairly intuitive. The weights for the observations depend mainly on the distance to the estimated point. The main idea behind the histogram to estimate the pdf is

$$\hat{f}_h(x) \approx \frac{\hat{F}(x + h) - \hat{F}(x - h)}{2h}, \tag{5.3}$$

where $\hat{F}$ is the ecdf. If h is small, the approximation method works well, producing smaller bin widths and a smaller bias. Rearranging (5.3) yields

$$\hat{f}_h(x) = \frac{1}{2nh} \sum_{i=1}^n I(x + h \geq x_i > x - h).$$

Different weights for observations in the vicinity of x can be achieved by simply multiplying $I\{x + h \geq x_i > x - h\}$ with the desired weight w. The kernel for the histogram is defined as

$$K(x - x_i) = \frac{1}{h} I(x + h \geq x_i > x - h) w(x_i),$$

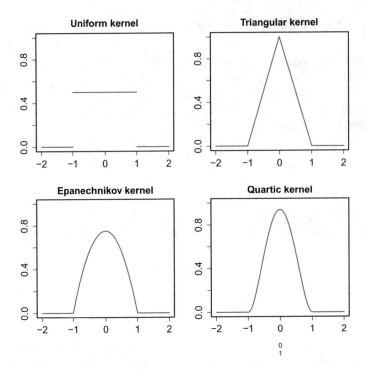

Fig. 5.5 Popular kernel functions. 🔍 BCS_PopularKernels

Table 5.1 Popular kernels
and their weighting functions

Kernel	Weighting function
Epanechnikov	$\frac{3}{4\sqrt{5}}(1 - \frac{u^2}{5})I(|u| \le 1)$
Triangular	$1 - |u|I(|u| \le 1)$
Uniform	$\frac{1}{2}I(|u| \le 1)$
Quartic	$\frac{15}{16}(1 - u^2)^2 I(|u| \le 1)$
Gaussian	$\frac{1}{\sqrt{2\pi}} \exp\{-\frac{1}{2}u^2\}$

where $w(x_i)$ is the weight for observation x_i, and depends on the distance of x_i from x. The sum of the weights must be $\sum_{i=1}^{n} w(x_i) = 1$ and $|w(x_i)| \le 1$ for all i. The density estimator for the pdf is then

$$\hat{f}_h(x) = \frac{1}{h} \sum_{i=1}^{n} K(x - x_i).$$

$K(x - x_i)$ is the uniform kernel weighting function and smooths the histogram.

Fig. 5.6 Nonparametric
density estimation of the
temperature at New Haven.
⌀BCS_Kernel_nhTemp

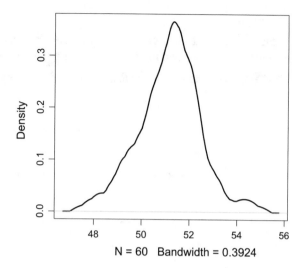

In R the simplest command to execute a kernel density estimation is `density()`.
One can implement all of the kernels mentioned above via the function argument
`kernel`. Returning to the temperature dataset let us estimate the kernel density for
the temperature in New Haven, Connecticut, using the Epanechnikov kernel. The
density is then estimated and plotted via (see Fig. 5.6)

```
> plot(density(nhtemp, kernel = "epanechnikov"))
```

To find the optimal bandwidth h for a kernel estimator, a similar problem has to
be solved as for the optimal binwidth. In practice one can use Silverman's rule of
thumb:

$$h^* = 1.06 \cdot \hat{\sigma} n^{-\frac{1}{5}}.$$

It is only a rule of thumb, because this h^* is only the optimal bandwidth under normality. But this bandwidth will be close to the optimal bandwidth for other distributions.
The optimal bandwidth depends on the kernel and the true density, see Härdle et al.
(2004).

5.1.5 Location Parameters

"Where are the data centered?" "How are the data scattered around the centre?" "Are
the data symmetric or skewed?" These questions are often raised when it comes to
a simple description of sample data. Location parameters describe the centre of a
distribution through a numerical value. They can be quantified in different ways and
visualised particularly well by boxplots.

Arithmetic mean

The term *arithmetic mean* characterises the average position of the realisations on the variable axis. It is a good location measure for data from a symmetric distribution.

Definition 5.2   The *sample (arithmetic) mean* for a sample of n values, $x_1, x_2, ..., x_n$ is defined by

$$\bar{x} = n^{-1} \sum_{i=1}^{n} x_i. \tag{5.4}$$

Applying the notions of absolute and relative frequencies, this formula can then be rewritten as

$$\bar{x} = n^{-1} \sum_{j=1}^{k} a_j n(a_j) = \sum_{j=1}^{k} a_j h(a_j).$$

According to the *Law of Large Numbers*, if $\{X_i\}_{i=1}^{n}$ denote n i.i.d. rvs with the same finite expected value $E(X_i) = \mu$, then their sample means converge almost surely to μ (also called the *population mean*), i.e.

$$n^{-1} \sum_{i=1}^{n} X_i \xrightarrow{a.s.} \mu \qquad \text{when } n \to \infty.$$

The arithmetic mean is calculated by mean().

```
> mean(nhtemp) # average temperature in New Haven
[1] 51.16
```

α-trimmed mean

The arithmetic mean is very often used as a location parameter, although it is not very robust, since its value is sensitive to the presence of outliers. In order to eliminate the outliers, one can trim the data by dropping a fraction $\alpha \in [0, 0.5)$ of the smallest and largest observations before calculating the arithmetic mean. This type of arithmetic mean, called the *α-trimmed mean*, is more robust to outliers. However, there is no unified recommendation regarding the choice of α. In order to define the trimmed mean, we need to define order statistics first.

Definition 5.3   Let $x_{(1)} \leq x_{(2)} \leq \ldots \leq x_{(n)}$ be the *sorted realizations* of the rv X. The term $x_{(i)}$, $i = 1, \ldots, n$ is called the ith *order statistic*, and in particular, $x_{(1)}$ is called the *sample minimum* and $x_{(n)}$ the *sample maximum*.

Definition 5.4   The *α-trimmed mean* is the arithmetic mean computed after trimming the fraction α of the smallest and largest observations of X, given by

$$\bar{x}_\alpha = \frac{1}{n - 2\lfloor n\alpha \rfloor} \sum_{i=\lfloor n\alpha \rfloor + 1}^{n - \lfloor n\alpha \rfloor} x_{(i)}$$

with $\alpha \in [0, 0.5)$. Where $\lfloor a \rfloor$ is the floor function, returning the largest integer not greater than a, see Sect. 1.4.1.

The argument `trim` is used in the function `mean` to compute the α-trimmed mean.

```
> mean(nhtemp, trim = 0.2)              > mean(nhtemp, trim = 0.4)
[1] 51.22222                            [1] 51.225
```

Quantiles

Another type of location parameter is the *quantile*. Quantiles are very robust, i.e. not influenced by outliers, since they are determined by the rank of the observations and they are estimates of the theoretical quantiles, see Definition 4.9.

Definition 5.5 The *p-quantile* $\tilde{x}_p$, where $0 \leq p \leq 1$, is a value such that at most $100 \cdot p\%$ of the observations are less than or equal to $\tilde{x}_p$ and $100 \cdot (1 - p)\%$ are greater than or equal to $\tilde{x}_p$. The number of observations which are less than or equal to $\tilde{x}_p$ is, then, equal to $\lceil np \rceil$.

$$\tilde{x}_p = \begin{cases} x_{(\lceil np \rceil)} & if\ np \notin \mathbb{Z} \\ \frac{1}{2}\left\{x_{(np)} + x_{(np+1)}\right\} & if\ np \in \mathbb{Z} \end{cases} \quad for\ p \in [0, 1].$$

Where $\lceil a \rceil$ is the ceiling function, returning the smallest integer not less than a, see Sect. 1.4.1. The sample quartiles are a special case of quantiles: the *lower quartile* $Q_1 = \tilde{x}_{0.25}$, the *median* $Q_2 = \tilde{x}_{0.5} = $ med, and *upper quartile* $Q_3 = \tilde{x}_{0.75}$. These three quartile values $Q_1 \leq Q_2 \leq Q_3$ divide the sorted observations into four segments, each of which contains roughly 25% of the observations in the sample.

To calculate the p-quantiles $\tilde{x}_p$ of the sample `nhtemp`, one uses `quantile()`. This function allows up to 9 different methods of computing the quantile, all of them converge asymptotically, as the sample size tends to infinity, to the true theoretical quantiles (`type = 2` is the method discussed here). Leaving the argument `probs` blank, R returns by default an array containing the 0, 0.25, 0.5, 0.75 and 1 quantiles, which are the sample minimum $x_{(1)}$, the lower quartile Q_1, the median Q_2 (or med), the upper quartile Q_3, and the sample maximum $x_{(n)}$. The median can be also found using `median()`.

```
> quantile(nhtemp, probs = 0.2)        # 20% quantile
20%
50.2
> median(nhtemp)                        # median
[1] 51.2
> quantile(nhtemp, probs = c(0.2, 0.5)) # 20% and 50% quantiles
 20%    50%
50.2 51.2
> quantile(nhtemp)                      # all quartiles
  0%    25%    50%    75%   100%
47.90 50.575 51.20 51.90 54.60
```

Mode

The mode is the most frequently occurring observation in a data set (also called the most fashionable observation). Together with the mean and median, one can use it as an indicator of the skewness of the data. In general, the mode is not equal to either the mean or the median, and the difference can be huge if the data are strongly skewed.

Definition 5.6 The *mode* is defined by

$$x_{mod} = a_j, \quad \text{with} \quad n(x_{mod}) > n(a_i), \quad \forall i \in \{1, ..., k\},$$

where $n(x)$ is the absolute frequency of x.

Note that the mode is not uniquely defined if several observations have the same maximal absolute frequency. There is no function in R that directly finds the sample mode of given data. The sample mode can be calculated using the command `names(sort(table(x), decreasing = TRUE))[1])`, where x is a vector. Let us consider again the dataset `nhtemp`.

```
> as.numeric(names(sort(table(nhtemp), decreasing = TRUE))[1])
[1] 50.9
```

These nested functions are better understood from the inside out. The function `table()` creates a frequency table for the observations in the dataset `nhtemp`, calculating the frequency for every single value. `sort()` with the argument `decreasing = TRUE` sorts the frequency table in decreasing order, so that the element with the highest frequency, i.e. the mode, appears first. Its name, the unique value for which the frequency was calculated, is then extracted by the function `names()[1]`, where `[1]` restricts the output of `names()` to the first position of the vector. Lastly, as the result is a string, here '50.9', it needs to be converted into a number by `as.numeric()`.

If it is desired to have the usual location parameters, such as the median, mean and some quantiles at once, one can use the command `summary()`.

```
> summary(nhtemp)
   Min. 1st Qu.  Median    Mean 3rd Qu.    Max.
  47.90   50.58   51.20   51.16   51.90   54.60
```

In general, `summary()` also produces summaries of the results of model fitting functions. Depending on the `class` of the first argument, particular methods are employed for this function. For example, the functions `summary.lm()` and `summary.glm()` are particular methods used to summarise the results produced by `lm` and `glm`, see Sect. 7.2.

5.1.6 Dispersion Parameters

One characteristic of a set of observations described by the measures of location is the typical or central value. However, it is also interesting to know how dispersed the observations are. Evaluating measures of dispersion in addition to measures of location provides a more complete description of the data.

Total range

Definition 5.7 The *total range* is the difference between the sample maximum and the sample minimum, i.e.

$$\text{TotalRange} = x_{(n)} - x_{(1)}.$$

Since the total range depends only on two observations, it is very sensitive to outliers and is thus a very weak dispersion parameter.

The function `range()` returns an array containing two values, namely the sample minimum and maximum. `diff()` calculates the difference between values by subtracting each value in a vector from the subsequent value. To obtain the total range, one simply calculates the first difference of the array given by the function `range()` using `diff()`.

```
> range(nhtemp)                      # sample min and sample max
[1] 47.9 54.6
> totalrange = diff(range(nhtemp)) # difference between max and min
> totalrange
[1] 6.7
```

Interquartile range

Definition 5.8 The *interquartile range* (IQR) of a sample is the difference between the upper quartile $\tilde{x}_{0.75}$ and the lower quartile $\tilde{x}_{0.25}$, i.e.

$$\text{IQR} = \tilde{x}_{0.75} - \tilde{x}_{0.25}.$$

It is also called the *midspread* or *middle fifty*, since roughly fifty percent of the observations are found within this range. The IQR is a robust statistic and is therefore preferred to the total range.

Currently, there is no function in R which directly gives the IQR. To find the upper and lower quartiles, one uses the function `quantile()` with `probs = c(0.25, 0.75)`, meaning that R should return the 0.25-quantile and the 0.75-quantile. In this example, the function `diff()` computes the IQR, i.e. the difference between the lower and upper quantiles.

```
> LUQ = quantile(nhtemp, probs = c(0.25, 0.75)); LUQ
   25%     75%
50.575 51.900
> IQR = diff(LUQ); IQR
   75%
1.325
```

Variance

The variance is one of the most widely used measures of dispersion. The variance is sensitive to outliers and is only reasonable for symmetric data.

Definition 5.9 The *sample variance* for a sample of n values $x_1, x_2, \ldots, x_n$ is the average of the squared deviations from the sample mean $\bar{x}$:

$$\tilde{s}^2 = \frac{1}{n} \sum_{i=1}^{n} (x_i - \bar{x})^2. \tag{5.5}$$

The unbiased *variance estimator* also called the *empirical variance* for a sample of n values $x_1, x_2, \ldots, x_n$ is the sum of the squared deviations from their mean $\bar{x}$ divided by $n - 1$, i.e.

$$s^2 = \frac{n}{n-1}\tilde{s}^2 = \frac{1}{n-1} \sum_{i=1}^{n} (x_i - \bar{x})^2. \tag{5.6}$$

Having copies $X_1, \ldots, X_n$ of a rv $X \sim (\mu, \sigma^2)$, it can be shown that

$$\mathsf{E}(S^2) = \mathsf{E}\frac{1}{n-1}\left\{ \sum_{i=1}^{n} (X_i - \bar{x})^2 \right\} = \sigma^2,$$

which means that S^2 is an unbiased estimator of $\mathsf{Var}(X) = \sigma^2$.

Standard deviation

As the variance is, due to the squares, not on the same scale as the original data, it is useful to introduce a normalised dispersion measure.

Definition 5.10 The *sample standard deviation* $\tilde{s}$ and the estimator for the population standard deviation based on the unbiased variance estimator are calculated from (5.5) and (5.6):

$$\tilde{s} = \sqrt{\tilde{s}^2}, \qquad s = \sqrt{s^2}.$$

The R functions var() and sd() compute estimates for the variance and standard deviation using the formulas for the unbiased estimators s^2 and s.

```
> var(nhtemp)            > sd(nhtemp)
[1] 1.601763             [1] 1.265608
```

Median absolute deviation

When computing the standard deviation, the distances to the mean are squared, thereby assigning more weight to large deviations. The standard deviation is thus very sensitive to outliers. Alternatively, one can use a more robust measure of dispersion, the median absolute deviation. It is robust since the median is less sensitive to outliers and the distances are not squared, effectively reducing the weight of outliers.

Definition 5.11 The *median absolute deviation* (MAD) is the median of the absolute deviations from the median:

$$MAD = \text{med } |x_i - \tilde{x}_{0.5}|, \qquad \forall i \in \{1, \ldots, n\}.$$

The function mad() returns by default the MAD according to Definition 5.11. However, if it is desired to compute the median of the absolute deviation from some other values, one simply includes the argument center. Below is an example of deviations both from the median and from the mean, that uses measurements of the annual flow of the river Nile at Ashwan between 1871 and 1970 (discharge in 10^8 m^3).

```
> mad(Nile)                          > mad(Nile, center = mean(Nile))
[1] 179.3946                         [1] 178.6533
```

Besides the dispersion measures discussed above, an alternative robust approach would be to measure the average absolute distance of each realisation from the median or the mean, i.e.

$$d_1 = \frac{1}{n} \sum_{i=1}^{n} |x_i - \tilde{x}_{0.5}| \quad \text{or} \quad d_2 = \frac{1}{n} \sum_{i=1}^{n} |x_i - \bar{x}|.$$

5.1.7 Higher Moments

The sample estimate of the skewness $S(X)$, see Definition 4.5, is given through

$$\hat{S} = \frac{\frac{1}{n} \sum_{i=1}^{n} (x_i - \bar{x})^3}{\{\frac{1}{n-1} \sum_{i=1}^{n} (x_i - \bar{x})^2\}^{3/2}},$$

and the sample excess kurtosis, see Definition 4.6, is provided by

$$\hat{K} = \frac{\frac{1}{n} \sum_{i=1}^{n} (x_i - \bar{x})^4}{\{\frac{1}{n} \sum_{i=1}^{n} (x_i - \bar{x})^2\}^2} - 3$$

which is implemented in R in package moments in functions skewness and kurtosis respectively.

```
> require(moments)              >
> skewness(Nile)                > kurtosis(Nile) - 3
[1] 0.3223697                   [1] -0.3049068
```

5.1.8   Box-Plot

The *box-plot* (or box-whisker plot) is a diagram which describes the distribution of a given data set. It summarises the location and dispersion measures discussed previously. The box-plot gives a quick glimpse of the observations' range and empirical distribution.

This box-plot of the dataset Nile visualizes the skewness of the data very well, see Fig. 5.7. Since the median, shown by the middle line, is not in the centre of the box, the data are not symmetric and the results for calculating the median absolute deviation from the median or from the mean differ, as we have just shown in the code above.

Let us now analyse the dataset nhtemp using the command boxplot(). The output is given in Fig. 5.8.

```
> boxplot(nhtemp)
```

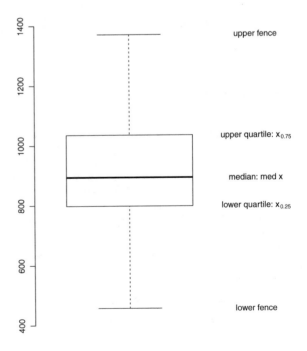

Fig. 5.7 Box-plot of the data Nile. **Q** BCS_Boxplot2

Fig. 5.8 Box-plot of the data nhtemp. **Q** BCS_Boxplot

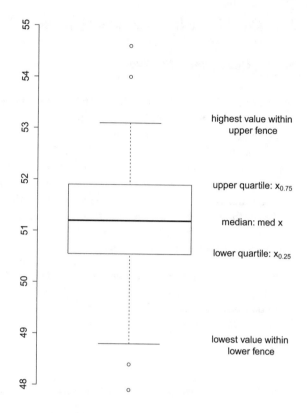

Approximately fifty percent of the observations are contained in the box. The upper edge is the 0.75-quantile and the lower edge is the 0.25-quantile. The distance between these two edges is the interquartile range (IQR). The median is indicated by the horizontal line between the two edges. If its distance from the upper edge is not equal to its distance from the lower edge, then the data are skewed.

The vertical lines extending outside the box are called *whiskers*. In the absence of outliers, the ends of the whiskers indicate the sample maximum and minimum. Otherwise, the ends of the whiskers lie at the highest value that is still within the upper fence ($\tilde{x}_{0.75} + 1.5 \cdot IQR$) and the lowest value that is still within the lower fence ($\tilde{x}_{0.25} - 1.5 \cdot IQR$). The factor 1.5, used by default, can be modified by setting the argument `range` appropriately. The (suspected) outliers are denoted by the points outside the two fences. For R not to draw the outliers, we set the argument `outline = FALSE`.

Another way of producing a box-plot is using the package `lattice` (more details in Chap. 10). The function used here is `bwplot()`. Consider again the dataset nhtemp. Since nhtemp is a time series object, it is converted to a vector using `as.vector()` in order for `bwplot` to work for the data nhtemp.

```
> require(lattice)
> bwplot(as.vector(nhtemp))
```

5.2   Confidence Intervals and Hypothesis Testing

5.2.1   Confidence Intervals

When estimating a population parameter θ (e.g. the population mean μ or the variance σ^2), it is important to have some clue about the precision of the estimation. The precision in this context is the probability that the estimate $\hat{\theta}$ is wrong by less than a given amount. It can be calculated using a random sample of size n drawn from the population. For most cases, like $\theta = \mu$, the sample size n must be large enough so that $\hat{\theta}$ can be assumed to be normally distributed (*Central Limit Theorem* 6.5).

The standard error of the sample mean measures the accuracy of the estimation of the mean, and the confidence interval quantifies how close the sample mean is expected to be to the population mean. Furthermore, it is naturally desirable to have a confidence interval as short as possible, something which is induced by large samples.

Definition 5.12   The *confidence interval (CI)* for the parameter θ of a continuous rv is a range of feasible values for an unknown θ together with a confidence coefficient $(1 - \alpha)$ conveying one's confidence that the interval actually covers the true θ. Formally, it is written as

$$P(\theta \in CI) = 1 - \alpha.$$

Confidence intervals for μ when σ is known

Since the value of σ is unknown in most cases, calculating confidence intervals when σ is known is not the typical situation. However, it can sometimes be known from past data in repeated surveys, from surveys on similar populations, or from theoretical considerations.

Consider a normal population with an unknown mean μ and known standard deviation σ. Let $X_i \sim N(\mu, \sigma^2)$ for $i = 1, ..., n$ be the sample rv's. Then $\bar{X} = \frac{1}{n} \sum_{i=1}^{n} X_i \sim N(\mu, \sigma^2/n)$ and $\sqrt{n}\frac{\bar{X}-\mu}{\sigma} \sim N(0, 1)$.

Now, let $Z \sim N(0, 1)$ with such $z_{1-\frac{\alpha}{2}}$ that

$$P(-z_{1-\frac{\alpha}{2}} \le Z \le z_{1-\frac{\alpha}{2}}) = 1 - \alpha.$$

With $Z = \sqrt{n} \cdot \frac{\bar{X}-\mu}{\sigma}$, this implies

$$P\left(\bar{X} - z_{1-\frac{\alpha}{2}} \cdot \frac{\sigma}{\sqrt{n}} \le \mu \le \bar{X} + z_{1-\frac{\alpha}{2}} \cdot \frac{\sigma}{\sqrt{n}}\right) = 1 - \alpha.$$

Thus, for a fixed $\alpha \in [0, 1]$, the confidence coefficient is $(1-\alpha)$ and the corresponding $100 \cdot (1 - \alpha)\%$-*confidence interval* for the population mean μ, assuming that the population is normally distributed and σ is known (Fig. 5.9), is given by

Fig. 5.9 The
$100 \cdot (1 - \alpha)\%$-confidence
interval for Z is the interval
between $z_{\frac{\alpha}{2}}$ and $z_{1-\frac{\alpha}{2}}$. The
area under the normal
density within this interval is
equal to $1 - \alpha$.
Q BCS_Conf2sided

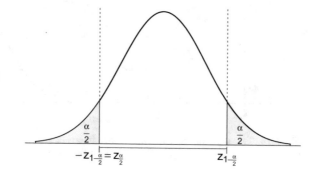

$$\left[\bar{x} - z_{1-\frac{\alpha}{2}} \cdot \frac{\sigma}{\sqrt{n}} ; \ \bar{x} + z_{1-\frac{\alpha}{2}} \cdot \frac{\sigma}{\sqrt{n}} \right].$$

Note that $z_{1-\frac{\alpha}{2}}$ is a $1 - \frac{\alpha}{2}$-quantile for a standard normal distribution, thus $-z_{1-\frac{\alpha}{2}} = z_{\frac{\alpha}{2}}$. To find the *lower* and *upper* limits ($z_{\frac{\alpha}{2}}$ and $z_{1-\frac{\alpha}{2}}$), one uses the quantile function for the normal distribution qnorm(), see Sect. 4.4.

These simple confidence intervals are best computed manually in R. For the example below, we again use the data nhtemp and assume that the standard deviation σ of 1.25 is known, for example.

```
> smean   = mean(nhtemp)             # sample mean
> n       = length(nhtemp)           # sample size
> alpha   = 0.1                      # confidence level
> sigma   = 1.25                     # sd known
> z       = qnorm(1 - alpha / 2)     # norm. distr. quantiles
> CI      = c(smean - z * sigma / sqrt(n), # confidence interval
+    smean + z * sigma / sqrt(n))
> CI
[1] 50.89456 51.42544
```

Sometimes only an upper limit or a lower limit for μ is desired, but not both. These are called *one-sided (one-tailed) confidence limits*. For example, a toy is considered to be harmful to children if it contains an amount of mercury that exceeds a certain value. A European buyer wants a guarantee from a European company that their products comply with European safety laws. The transaction may then take place if the 99%-confidence *upper limit* does not exceed the desired maximum. In the same contract, one does not want too many failures in the shipped good, e.g. the 95%-confidence *lower limit* should not exceed the desired minimum (Fig. 5.10). Taking $Z = \sqrt{n}\frac{\bar{X}-\mu}{\sigma} \sim N(0, 1)$, it follows that

$$P\left(\mu \geq \bar{X} - z_{1-\alpha} \cdot \frac{\sigma}{\sqrt{n}} \right) = \alpha.$$

Thus, the $100 \cdot \alpha\%$-confidence lower limit is $\bar{x} - z_{1-\alpha}\frac{\sigma}{\sqrt{n}}$ and the upper limit is given by $\bar{x} + z_{1-\alpha}\frac{\sigma}{\sqrt{n}}$.

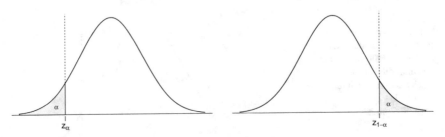

Fig. 5.10 Two types of one-sided confidence intervals: *lower* limit (*left*) and *upper* limit (*right*) confidence interval. Q BCS_Conf1Sidedleft, Q BCS_Conf1sidedright

Confidence intervals for μ when σ is unknown

Since the population's σ is generally unknown, the construction of confidence intervals for μ will usually be based on Student's t-distribution (see Sect. 4.4.2). Define $V \sim t_\nu$ (meaning that V is t-distributed with ν degrees of freedom) by

$$P(-t_{1-\frac{\alpha}{2},\nu} \leq V \leq t_{1-\frac{\alpha}{2},\nu}) = 1 - \alpha.$$

Now, assuming that the population is normally distributed, and the sample size is n, it follows that $V = \sqrt{n}\frac{\bar{X}-\mu}{S} = \sqrt{n}\frac{\frac{1}{n}\sum_{i=1}^{n}X_i-\mu}{\sqrt{\frac{1}{n-1}\{\sum_{i=1}^{n}(X_i-\bar{X})^2\}}}$, thus

$$P\left(\bar{X} - t_{1-\frac{\alpha}{2},n-1} \cdot \frac{S}{\sqrt{n}} \leq \mu \leq \bar{X} + t_{1-\frac{\alpha}{2},n-1} \cdot \frac{S}{\sqrt{n}}\right) = 1 - \alpha.$$

Definition 5.13 The $100 \cdot (1-\alpha)\%$-confidence interval for the population mean μ when the population is normally distributed and σ is unknown is defined by

$$\left[\bar{x} - t_{1-\frac{\alpha}{2},n-1} \cdot \frac{S}{\sqrt{n}}; \bar{x} + t_{1-\frac{\alpha}{2},n-1} \cdot \frac{S}{\sqrt{n}}\right].$$

When calculating the confidence interval for σ unknown, the R code from above changes only a little. To find $t_{1-\alpha/2,n-1}$, we use the function qt (), see again Sect. 4.4.

```
> smean   = mean(nhtemp)
> sigma2 = sd(nhtemp)                           # estimate sd from data
> n       = length(nhtemp)                       # sample size
> alpha  = 0.1                                    # confidence level
> z2      = qt(1 - alpha / 2, n - 1)            # the t-distr. quantiles
> CI2     = c(smean - z2 * sigma2 / sqrt(n),   # confidence interval
+      smean + z2 * sigma2 / sqrt(n))
> CI2
[1] 50.88696 51.43304
```

Note that this confidence interval is slightly larger than the one calculated for σ known. In other words, having to estimate the variance introduces more uncertainty into our estimation.

5.2.2 Hypothesis Testing

A hypothesis test, or *test of significance*, is a widely used tool in statistical analysis. Based on the information gained from a sample, one attempts to make a decision about a hypothesis. The hypotheses in this case are assumptions about the parameters of the distribution followed by the rv in the population (e.g. the mean μ, proportion p, variance σ^2, standard deviation σ, etc.). Similarly to parameter estimation, where no estimates are exactly equal to the parameters, it can never be concluded with certainty whether a hypothesis is wrong or correct. Therefore, one can only *reject* or *fail to reject* a hypothesis.

When conducting a test, two hypotheses are made, namely the null hypothesis H_0, and an alternative hypothesis H_1 (or often also H_a). Suppose a null hypothesis about a parameter θ is made: one desires to know whether θ is equal to a certain value, which is denoted by θ_0. Then the alternative hypothesis H_1 is that θ is not equal to θ_0:

$$H_0 : \theta = \theta_0 \quad \text{vs} \quad H_1 : \theta \neq \theta_0.$$

It is important to note that the hypotheses are mutually exclusive. The test above is called a *two-sided* or *two-tailed* test, since the alternative hypothesis H_1 does not make any reference to the sign of the difference $\theta - \theta_0$. Therefore, the interest here lies only in the absolute values of $\theta - \theta_0$.

However, sometimes the investigator notes only deviations from the null hypothesis H_0 in one direction and ignores deviations in other directions. The investigators could, for example, be certain that if θ is not less than or equal to θ_0, then θ must be greater than θ_0 or vice versa. Formally:

$$H_0 : \theta \leq \theta_0 \quad \text{vs} \quad H_1 : \theta > \theta_0,$$

$$\text{or}$$

$$H_0 : \theta \geq \theta_0 \quad \text{vs} \quad H_1 : \theta < \theta_0.$$

Each time when conducting a statistical test, one faces two types of risk:

Type I error is the error of rejecting a null hypothesis when it is in fact true.
Type II error is the error of failing to reject a null hypothesis when it is actually not true.

These two types of risks are treated in different ways: it is always desired to have the probability of type I error, denoted by α, be as small as possible. On the other hand, since it is ideal that a test of significance rejects a null hypothesis when it is false, it is desired to have the probability of type II error, denoted by β, as small as possible too.

Test for the mean of a normal population (σ known)

Suppose $X \sim N(\mu, \sigma^2)$. For simplicity, it is assumed that σ is known. A random sample $\{x_i\}$ of size n is drawn from X. The null hypothesis of the two-sided test for the mean is then defined as follows: $H_0 : \mu = \mu_0$. Using the information in the sample, we have to make a decision about the null hypothesis H_0: should it be rejected or not? In most cases, one looks at the deviation of the sample mean $\bar{x}$ from the hypothetical value μ_0:

$$|\bar{x} - \mu_0| > 0.$$

However the null hypothesis H_0 can not be immediately rejected. Deviations may occur even if H_0 is true, for example through an unfavorable sampling. Only when it exceeds a certain critical value is the deviation said to be *statistically significant*, or simply *significant*, and therefore one rejects the null hypothesis H_0.

   The question is now: how to decide whether or not the deviation is significant? We fail to reject the null hypothesis H_0 as long as the estimated confidence interval for μ of the same sample contains the hypothetical value μ_0 (this constitutes a connection between hypothesis testing and confidence intervals).

The Critical Region The critical region (or the region of rejection) is the set of values of $\bar{x}$ that cause the rejection of the null hypothesis H_0. It can be determined with the distribution of the sample mean $\bar{x}$. The probability of a type I error (also called the *significance level* of a test), i.e. the probability of rejecting the null hypothesis H_0 although it is true, should be at most α:

$$P(\bar{x} \in \text{critical region} \mid H_0) = \alpha.$$

The construction of the critical region depends on the type of test one conducts (two-sided or one-sided).

Two-Sided Tests Recall the hypotheses

$$H_0 : \mu = \mu_0 \quad \text{vs} \quad H_1 : \mu \neq \mu_0.$$

In a test for the mean, a natural criterion for judging whether the observations favour H_0 or H_1 is *the size of the deviation of the sample mean $\bar{x}$ from the hypothetical value* μ_0, i.e. $(\bar{x} - \mu_0)$. Under the null hypothesis H_0, $(\bar{X} - \mu_0) \sim N(0, \sigma^2/n)$. Under the alternative hypothesis H_1, $(\bar{X} - \mu_0) \sim N(\mu_1 - \mu_0, \sigma^2/n)$, where $\mu_1 - \mu_0 \neq 0$.

   Large values of the *test criterion* $(\bar{x} - \mu_0)$ can cause the rejection of H_0 in favour of H_1. However there is no exact answer how large these values should be. The larger is the value of $(\bar{x} - \mu_0)$ required to reject H_0, the smaller is the probability of a type I error α, but the higher is the probability of a type II error β. Larger samples minimise both errors, but may be difficult to obtain and therefore inefficient. In practice, the critical value is determined so that the probability of type I errors α is 0.05. This is called a *test at the 5% level*. Sometimes a level of 1% is chosen when incorrect rejection of the null hypothesis H_0 is considered as a serious mistake.

The method of testing the hypothesis $\mu = \mu_0$ is as follows:

1. Transform $(\bar{X} - \mu_0)$ into under H_0 a standard normal variable

$$Z = \sqrt{n}\,\frac{\bar{X} - \mu_0}{\sigma} \sim N(0, 1).$$

Recall that in a two-sided test, only absolute values of $|Z|$ are relevant, since there is no reference to the sign of $(\bar{x} - \mu_0)$ in the alternative hypothesis H_1.

2. Reject the null hypothesis $H_0 : \mu = \mu_0$ if z as a realization of Z fulfills

$$z \in \text{critical region} = (-\infty, -z_{1-\frac{\alpha}{2}}) \cup (z_{1-\frac{\alpha}{2}}, +\infty),$$

that is to say $|z| > z_{1-\frac{\alpha}{2}}$. The value $z_{1-\frac{\alpha}{2}}$ is determined in such a way that

$$\Phi(z_{1-\frac{\alpha}{2}}) = 1 - \frac{\alpha}{2},$$

where Φ is the cdf of $N(0, 1)$. It is important to note that the critical region of a two-sided test is symmetric, with a probability of $\frac{\alpha}{2}$ on each side. Thus one rejects the null hypothesis $H_0 : \mu = \mu_0$ if

$$\bar{x} \in \text{critical region} \equiv \left(-\infty, \mu_0 - z_{1-\frac{\alpha}{2}} \cdot \frac{\sigma}{\sqrt{n}}\right) \cup \left(\mu_0 + z_{1-\frac{\alpha}{2}} \cdot \frac{\sigma}{\sqrt{n}}, +\infty\right).$$

It is easy to see the connection between the two-sided hypothesis testing $\mu = \mu_0$ and the $100 \cdot (1 - \alpha)\%$-confidence interval for μ. According to the rule of hypothesis rejection, $H_0 : \mu = \mu_0$ fails to be rejected by a two-sided test when

$$\bar{x} \in \left(\mu_0 - z_{1-\frac{\alpha}{2}} \cdot \frac{\sigma}{\sqrt{n}}, \mu_0 + z_{1-\frac{\alpha}{2}} \cdot \frac{\sigma}{\sqrt{n}}\right).$$

Rearranging this equation, one obtains

$$\mu_0 \in \left(\bar{x} - z_{1-\frac{\alpha}{2}} \cdot \frac{\sigma}{\sqrt{n}}, \bar{x} + z_{1-\frac{\alpha}{2}} \cdot \frac{\sigma}{\sqrt{n}}\right),$$

a two-sided $100 \cdot (1 - \alpha)\%$-confidence interval for μ. It contains *those values of μ_0 for μ that would lead to a failure of rejection of H_0 using a two-sided test on the sample mean $\bar{x}$.*

One-Sided Tests Unlike in a two-sided test, the critical region of a one-sided test is not symmetric. Thus, the hypotheses do not concern a single discrete value, but are expressed as

$$H_0 : \mu \le \mu_0 \quad \text{vs} \quad H_1 : \mu > \mu_0.$$

It is often needed when a new treatment (e.g. scholarships for disadvantaged students) is of no interest unless it is superior to the standard treatment (no scholarships). Thus, the null hypothesis can be expressed as H_0: the average grade of disadvantaged students does not increase after they receive scholarships. The rule is to reject H_0 when

$$z = \sqrt{n}\,\frac{\bar{x} - \mu_0}{\sigma} > z_{1-\alpha}.$$

The critical region is the interval $(z_{1-\alpha}, +\infty)$. The area under the bell curve $\varphi(\cdot)$ within this interval is α, too.

Inversely, when the investigator is interested to know whether or not the population mean is smaller than a certain value μ_0, the hypotheses are

$$H_0 : \mu \ge \mu_0 \quad \text{vs} \quad H_1 : \mu < \mu_0.$$

The rule to reject H_0 is

$$z = \sqrt{n}\,\frac{\bar{x} - \mu_0}{\sigma} < z_{\alpha}.$$

The critical region is the interval $(-\infty, z_{\alpha})$. The area under the bell curve $\varphi(\cdot)$ within this interval is α, too.

Manual Hypothesis Testing Unfortunately, there is no function in R that does hypothesis testing under the assumption that σ is known. But one can easily compute the statistics manually.

For the following example, consider nottem, a sample of size $n = 240$ containing average monthly air temperatures at Nottingham Castle in degrees Fahrenheit for the period 1920–1939, which is then converted to Celsius. Supposing the standard deviation in the population $\sigma = 5$, one desires to test whether the average monthly temperatures in degrees Celsius in Nottingham is equal to $\mu = 15.5$ using the sample mean $\bar{x} = 9.46$. The hypotheses are written as follows:

$$
\begin{aligned}
&H_0 : \mu = 15.5 \text{ vs } H_1 : \mu \neq 15.5 \text{ (for two-sided test)}, \\
&H_0 : \mu \ge 15.5 \text{ vs } H_1 : \mu < 15.5 \text{ (for left-sided test)}, \\
&H_0 : \mu \le 15.5 \text{ vs } H_1 : \mu > 15.5 \text{ (for right-sided test)}.
\end{aligned}
$$

The next testing step is to convert the test criterion into under H_0 a standard normal variable.

```
> nottemc = (nottem - 32) * 5 / 9      # Fahrenheit to Celsius
> n      = length(nottemc); n          # sample size
[1] 240
> z      = (mean(nottemc) - 15.5) / (5 / sqrt(n))
> z                                     # test statistics
[1] -18.69432
```

Depending on the type of test, one uses different decision rules for rejecting or not rejecting the null hypothesis H_0:

1. for the two-sided test, one rejects H_0 at $\alpha = 0.05$ since $|z| > z_{0.975} = 1.96$,
2. for the left-sided test, one rejects H_0 at $\alpha = 0.05$ since $z < z_{0.05} = -1.64$,
3. for the right-sided test, one cannot reject H_0 at $\alpha = 0.05$ since $z \not> z_{0.95} = 1.64$.

Test for the mean of a normal population (σ unknown)

Just as in the construction of confidence intervals, Student's t distribution can be used in hypothesis testing under certain circumstances, namely when the population's standard deviation σ is unknown and therefore needs to be estimated. An indispensable assumption for small samples is that the rv is approximately normally distributed. If this is the case, then the test statistic follows Student's t distribution. In large samples, this assumption is no longer obligatory since the sample means are approximately normally distributed by the *Central Limit Theorem*, see Serfling (1980). The decision rules for the hypothesis testing are similar to those of the test for the mean when σ is known, except that the standard normal distribution $N(0, 1)$ is replaced by the Student's t-distribution with $n - 1$ degrees of freedom, where n is the sample size.

Two-Sided Tests The rule for two-sided tests is to reject the null hypothesis $H_0 : \mu = \mu_0$ if

$$t = \sqrt{n}\, \frac{\bar{x} - \mu}{s} \in \text{critical region} \equiv (-\infty, -t_{n-1, 1-\frac{\alpha}{2}}) \cup (t_{n-1, 1-\frac{\alpha}{2}}, +\infty),$$

that is to say $t > t_{n-1, 1-\frac{\alpha}{2}}$. The value $t_{n-1, 1-\frac{\alpha}{2}}$ is determined so that

$$P(T \leq t_{n-1, 1-\frac{\alpha}{2}}) = 1 - \frac{\alpha}{2}, \qquad \text{where } T \sim t_{n-1}.$$

One-Sided Tests The rules for hypothesis testing in this case are analoguous to those used when σ is known.

For the hypotheses

$$H_0 : \mu \leq (\geq) \mu_0 \quad \text{vs} \quad H_1 : \mu > (<) \mu_0,$$

H_0 is rejected if

$$t = \sqrt{n}\,\frac{\bar{x} - \mu_0}{s} > (<)\, t_{n-1,\,1-\alpha}.$$

Hypothesis Testing Using p-Values The critical regions depend on the distribution of the test statistics and on the probability of a type I error α. This makes manual testing inconvenient, since for every new value of α the critical region has to be recomputed. To overcome this problem most of the tests in R can be performed using the concept of the p-value.

Definition 5.14 The p-value is the probability of obtaining a test criterion at least as large as the observed one, assuming that the null hypothesis is true.

For continuous distributions of the test criterion, the p-value can be determined as that value of α^* such that the test criterion coincides with the next boundary of the critical region. For the one-sided test for the mean with $H_0 : \mu \leq \mu_0$, this implies

$$\sqrt{n}\,\frac{\bar{x} - \mu_0}{\sigma} = z_{1-\alpha^*}.$$

Solving for α^* leads to

$$p\text{-value} = 1 - \Phi\left(\sqrt{n}\,\frac{\bar{x} - \mu_0}{\sigma}\right).$$

Similarly, for the one-sided test with $H_0 : \mu \geq \mu_0$, we obtain

$$p\text{-value} = \Phi\left(\sqrt{n}\,\frac{\bar{x} - \mu_0}{\sigma}\right).$$

In the case of the two-sided test, it can be shown using a similar logic that

$$p\text{-value} = 2 - 2\Phi\left(\left|\sqrt{n}\,\frac{\bar{x} - \mu_0}{\sigma}\right|\right).$$

In the case of unknown σ, the cdf Φ is replaced with the cdf of the t-distribution with n-1 degrees of freedom. For more complicated tests and distributions, the p-value should be determined individually.

The decision regarding the rejection of the null hypothesis is made using a simple scheme:

- the null hypothesis is rejected if the p-value is smaller than the prespecified significance level α;
- the null hypothesis is not rejected if the p-value is equal to or larger than α.

This decision rule is independent of the type of the test and the distribution of the test criterion, allowing for quick testing with different levels of α.

Using t.test() Hypothesis testing of the mean with unknown variance involving Student's t test can be used in R through the function t.test(). For the next example, consider the dataset nhtemp, a sample of size $n = 60$ containing the mean annual temperature x in degrees Fahrenheit in New Haven, Connecticut, with a sample mean of $\bar{x} = 51.16$. Suppose one wants to test whether the population mean μ is equal to the hypothetical value μ_0, say 50. This is a two-sided test of the mean:

$$H_0 : \mu = 50 \quad \text{vs} \quad H_1 : \mu \neq 50.$$

Under the assumption that the standard deviation σ is unknown, one uses t.test().

```
> t.test(x = nhtemp,
+    alternative = "two.sided",  # two-sided test
+    mu          = 50,           # for mu = 50
+    conf.level  = 0.95)         # at level 0.95

          One Sample t-test

data:  nhtemp
t = 7.0996, df = 59, p-value = 1.835e-09
alternative hypothesis: true mean is not equal to 50
95 percent confidence interval:
 50.83306 51.48694
sample estimates:
mean of x
 51.16
```

As can be seen from the listing above, beside the test statistics, the function t.test() returns the confidence intervals and the sample estimates. The hypothesis testing above leads to the rejection of the null hypothesis $H_0 : \mu = 10$ at 95%-confidence level. Obviously in this two-sided test, the hypothetical value $\mu_0 = 10$ lies outside the 95%-confidence interval. The absolute value of Student's t statistic is greater than the critical value $t_{59, 1-\frac{0.05}{2}}$, since the p-value is much smaller than $\alpha = 5\%$.

To conduct a one-sided test, the argument alternative must be changed into less or greater.

```
> t.test(x = nhtemp,
+    alternative = "less",   # one-sided test
+    mu          = 50,       # for mu < 50
+    conf.level  = 0.95)     # at level 0.95

          One Sample t-test

data:  nhtemp
t = 7.0996, df = 59, p-value = 1
alternative hypothesis: true mean is less than 50
95 percent confidence interval:
    -Inf 51.43304
sample estimates:
mean of x
 51.16
```

Thus the hypothesis testing with a hypothetical value $\mu_0 = 50$ leads to the rejection of $H_0 : \mu > \mu_0$ at the 95%-confidence level.

Testing σ^2 of a normal population

It is also interesting to see whether the population variance has a certain value of σ_0^2. The question is then: how to construct confidence intervals for σ^2 from the estimator s^2? How to test hypotheses about the value of σ^2? With a little modification of s^2, one can answer these questions by looking at a rv that follows a χ_ν^2 distribution.

If $X_1, ..., X_n$ are i.i.d. random normal variables, then using Definition 4.13 of the χ^2 distribution we obtain

$$\frac{(n-1)S^2}{\sigma^2} \sim \chi_{n-1}^2, \quad \text{with } S^2 = \frac{1}{n-1} \sum_{i=1}^n (X_i - \bar{X})^2.$$

Confidence Intervals for σ^2 Let $Y \sim \chi_\nu^2$. Now, choose $\chi_{\nu,1-\alpha/2}^2$ such that

$$P(Y \geq \chi_{\nu,1-\alpha/2}^2) = \alpha/2.$$

and $\chi_{\nu,\alpha/2}^2$ such that

$$P(Y \leq \chi_{\nu,\alpha/2}^2) = \alpha/2.$$

Since $Y = \frac{\nu \cdot S^2}{\sigma^2}$, it is easy to show that

$$P\left(\chi_{\nu,\frac{\alpha}{2}}^2 < \frac{\nu S^2}{\sigma^2} < \chi_{\nu,1-\frac{\alpha}{2}}^2\right) = 1 - \alpha,$$

which is equivalent to

$$P\left(\frac{\nu S^2}{\chi_{\nu,1-\frac{\alpha}{2}}^2} < \sigma^2 < \frac{\nu S^2}{\chi_{\nu,\frac{\alpha}{2}}^2}\right) = 1 - \alpha.$$

This is the general formula for a two-sided $100 \cdot (1 - \alpha)$%-confidence limit for σ^2. The number of degrees of freedom is $\nu = n - 1$ if s^2 is computed from a sample of size n.

Testing for σ^2 This situation occurs, for example, when a theoretical value of σ^2 is to be tested or when the sample data are being compared to a population whose σ^2 is known. If the hypotheses are

$$H_0 : \sigma^2 \leq (\geq) \sigma_0^2 \quad \text{vs} \quad H_1 : \sigma^2 > (<) \sigma_0^2,$$

then reject H_0 if

$$y = \frac{vs^2}{\sigma_0^2} = \frac{\sum_{i=1}^{n}(x_i\bar{x})^2}{\sigma_0^2} > \chi_{v,\alpha}^2 \left(< \chi_{v,1-\alpha}^2 \right).$$

For a two-sided test, reject the null hypothesis H_0 if

$$y < \chi_{v,1-\frac{\alpha}{2}}^2 \quad \text{or} \quad y > \chi_{v,\frac{\alpha}{2}}^2.$$

Note that for values $v > 100$, an approximation can be made by

$$Z = \sqrt{2Y} - \sqrt{2v-1} \sim N(0,1).$$

The rejection rule of the null hypothesis H_0 using this proxy variable is analogous to the test of a mean when σ is known.

Test for equal means $\mu_1 = \mu_2$ of two independent samples

In comparative studies, one is interested in the differences between effects rather than the effects themselves. For instance, it is not the absolute level of sugar concentration in blood reported for two types of diabetes medication that is of interest, but rather the difference between the levels of sugar concentration. One of many aspects of comparative studies is comparing the means of two different populations.

Consider two samples $\{x_{i,1}\}_{i\in\{1,...,n_1\}}$ and $\{x_{j,2}\}_{j\in\{1,...,n_2\}}$, independently drawn from $N(\mu_1, \sigma_1^2)$ and $N(\mu_2, \sigma_2^2)$ respectively. The two-sample test for the mean is as follows:

$$H_0 : \mu_1 - \mu_2 = \delta_0 \quad \text{vs} \quad H_1 : \mu_1 - \mu_2 \neq \delta_0.$$

There are two cases to distinguish:

1. both populations have the same standard deviation ($\sigma_1 = \sigma_2$);
2. the two populations have different standard deviations ($\sigma_1 \neq \sigma_2$).

Before testing the hypotheses, it is important to aquire information about the variance of the difference of the sample means $\sigma_{\bar{x}_1-\bar{x}_2}^2$.

Definition 5.15 Under the assumption of independent rvs, the variance of the difference between the sample means is defined as

$$\sigma_{\bar{x}_1-\bar{x}_2}^2 = \frac{\sigma_1^2}{n_1} + \frac{\sigma_2^2}{n_2}.$$

Furthermore, this variance can be estimated and used to construct the test statistics later on. The estimation of $\sigma_{\bar{x}_1-\bar{x}_2}^2$ depends on the assumptions about σ_1 and σ_2.

1. When both populations have the same variance $\sigma^2 = \sigma_1^2 = \sigma_2^2$, then σ^2 is estimated by the unbiased pooled estimator of the $\bar{x}_1 - \bar{x}_2$ variance s_{pooled}^2:

$$s_{pooled}^2 = \frac{\text{pooled sum of squares}}{\text{pooled degrees of freedom}} = \frac{(n_1 - 1)s_1^2 + (n_2 - 1)s_2^2}{n_1 + n_2 - 2}. \qquad (5.7)$$

Thus, the sample estimate $s_{\bar{x}_1 - \bar{x}_2}^2$ of the population variance $\sigma_{\bar{x}_1 - \bar{x}_2}^2$ is

$$s_{\bar{x}_1 - \bar{x}_2}^2 = s_{pooled}^2 \left(\frac{1}{n_1} + \frac{1}{n_2} \right).$$

2. When the populations have different variances $\sigma_1^2 \neq \sigma_2^2$, then $\sigma_{\bar{x}_1 - \bar{x}_2}^2$ is estimated by the following unbiased estimator:

$$s_{\bar{x}_1 - \bar{x}_2}^2 = \frac{s_1^2}{n_1} + \frac{s_2^2}{n_2}. \qquad (5.8)$$

Whether the first case applies can be investigated by the function var.test(), which uses the F-distribution introduced in Sect. 4.4.3. Consider in this example sleep, a data frame with 20 observations on 2 variables: the amount of extra sleep after taking a drug (extra) and the control group (group).

```
> # test for equal variances
> var.test(sleep$extra, sleep$group,
+    ratio       = 1,            # hypothesized ratio of variances
+    alternative = "two.sided",  # two-sided test
+    conf.level  = 0.95)         # at level 0.95

    F test to compare two variances

data:  sleep$extra and sleep$group
F = 15.4736, num df = 19, denom df = 19, p-value = 1.617e-07
alternative hypothesis: true ratio of variances is not equal to 1
95 percent confidence interval:
  6.124639 39.093291
sample estimates:
ratio of variances
        15.4736
```

The null hypothesis of equal variances of the groups is rejected. This result will be useful when testing for equal means.

Testing the Hypothesis $\mu_1 = \mu_2$ when $\sigma^2 = \sigma_1^2 = \sigma_2^2$

Under this assumption, use (5.7) for the estimator $s_{\bar{x}_1 - \bar{x}_2}$ as follows:

$$s_{\bar{x}_1 - \bar{x}_2} = \sqrt{s_{pooled}^2 \left(\frac{1}{n_1} + \frac{1}{n_2} \right)}.$$

Hence, the rejection rule for H_0 in the two-sided test

$$H_0 : \mu_1 = \mu_2 \quad \text{vs} \quad H_1 : \mu_1 \neq \mu_2,$$

is

$$|t| = \left| \frac{\bar{x}_1 - \bar{x}_2}{s_{\bar{x}_1 - \bar{x}_2}} \right| > t_{n_1 + n_2 - 2, \, 1 - \frac{\alpha}{2}}.$$

If the hypotheses are

$$H_0 : \mu_1 \leq (\geq) \, \mu_2 \quad \text{vs} \quad H_1 : \mu_1 > (<) \, \mu_2,$$

then reject H_0 if

$$t = \frac{\bar{x}_1 - \bar{x}_2}{s_{\bar{x}_1 - \bar{x}_2}} > t_{n_1 + n_2 - 2, \, 1 - \alpha} \quad \left(< -t_{n_1 + n_2 - 2, \, 1 - \alpha} \right).$$

Testing the Hypothesis $\mu_1 = \mu_2$ when $\sigma_1^2 \neq \sigma_2^2$

Under this assumption, use (5.8) and the estimator $s_{\bar{x}_1 - \bar{x}_2}$ as follows:

$$s_{\bar{x}_1 - \bar{x}_2} = \sqrt{\frac{s_1^2}{n_1} + \frac{s_2^2}{n_2}}.$$

Hence, the rejection rule for H_0 in two-sided tests

$$H_0 : \mu_1 = \mu_2 \quad \text{vs} \quad H_1 : \mu_1 \neq \mu_2,$$

is $|t| = \left| \frac{\bar{x}_1 - \bar{x}_2}{s_{\bar{x}_1 - \bar{x}_2}} \right| > t_{\nu, \, 1 - \frac{\alpha}{2}}.$
If the hypotheses are

$$H_0 : \mu_1 \leq (\geq) \, \mu_2 \quad \text{vs} \quad H_1 : \mu_1 > (<) \, \mu_2,$$

then reject H_0 if

$$t = \frac{\bar{x}_1 - \bar{x}_2}{s_{\bar{x}_1 - \bar{x}_2}} > t_{\nu, \, 1 - \alpha} \quad \left(< -t_{\nu, \, 1 - \alpha} \right).$$

The degrees of freedom ν are computed as

$$\nu = \left(\frac{s_1^2}{n_1} + \frac{s_2^2}{n_2} \right)^2 \left\{ \frac{(s_1^2/n_1)^2}{n_1 - 1} + \frac{(s_2^2/n_2)^2}{n_2 - 1} \right\}^{-1}.$$

Using `oneway.test()` Testing for equal means is done using the function `oneway.test`. The assumption about the variances can be specified in the argument `var.equal`. Consider again the dataframe `sleep`. Suppose we want to test whether the mean of the hours of sleep in the first group is equal to that of the second group ($H_0 : \mu_1 = \mu_2$).

```
> # assuming not equal variances
> oneway.test(extra ~ group, data = sleep, var.equal = FALSE)

        One-way analysis of means (not assuming equal variances)

data:  extra and group
F = 3.4626, num df = 1.000, denom df = 17.776, p-value = 0.0794
```

The test in R relies on the squared test criterion T^2, which follows an F-distribution with 1 and ν degrees of freedom. Using the p-value approach, one cannot reject the null hypothesis H_0 of equality of means of both groups at the 5%-level, since the p-value > 0.05. This applies in both cases, whether the variances are equal or not. However at the 10%-level, one rejects H_0 since the p-value < 0.1.

5.3 Goodness-of-Fit Tests

In the following, let F and G be two continuous distributions and $\{x_1, \ldots, x_n\}$ be a random sample from an rv X with unknown distribution. A common question is, what distribution does X have. Another frequently asked question is whether the observations $\{x_1, \ldots, x_n\}$ and $\{y_1, \ldots, y_m\}$ are realizations of rvs X and Y with the same distributions $F = G$. Table 5.2 gives an overview of the tests introduced later.

Table 5.2 Conducting nonparametric tests in R

Test	Samples	Scale requirement	R syntax	Null hypothesis
Kolmogorov–Smirnov	≤ 2	Interval	`ks.test()`	$F = G$
Anderson–Darling	≤ 2	Interval	`ad.test()`	$F = G$
Cramér–von Mises	≤ 2	Interval	`cvm.test()`	$F = G$
Shapiro–Wilk	≤ 2	Interval	`shapiro.test()`	$X \sim N$
Wilcoxon signed rank	≤ 2	Ordinal and paired	`wilcox.test()`	$\tilde{x}_{0.5} = c$
Mann–Whitney U	≤ 2	Ordinal and non-paired	`wilcox.test()`	$F_1(x) = F_2(x)$
Kruskal–Wallis	any	Ordinal	`Kruskal.test()`	$F_1(x) = \cdots = F_m(x)$

5.3.1 General Tests

The Kolmogorov–Smirnov, Anderson–Darling and Cramér–von Mises tests measure the distance between two possible distributions. These tests test whether a sample is the realization of a rv that follows some prespecified but arbitrary distribution. If the distance is large enough, the distributions are regarded as different. In R, the package `stats` implements the Kolmogorov–Smirnov test and the package `goftest` implements the Anderson–Darling and Cramér–von Mises tests.

Kolmogorov–Smirnov test

The Kolmogorov–Smirnov test determines whether two distributions F and G are significantly different:

$$H_0 : F = G \quad vs. \quad H_1 : F \neq G.$$

The underlying idea of this test is to measure the so-called Kolmogorov–Smirnov distance between two distributions, under the restriction that both functions are continuous. The measure for the distance is the supremum of the absolute difference between the two distribution functions:

$$KS = \sup_x |F(x) - G(x)|. \tag{5.9}$$

This test is commonly used to compare the ecdf to the assumed parametric one. The following example shows whether the standardised log-returns $\tilde{r}_{t,DAX} = \frac{r_{t,DAX} - \bar{r}_{DAX}}{s_{r_{DAX}}}$ of the DAX index follow a t-distribution with k degrees of freedom, where $r_{t,DAX} = \log \frac{P_{t,DAX}}{P_{t+1,DAX}}$ are log-returns, and $P_{t,DAX}$ prices at time t. Standardised log-returns have a zero mean and a unit standard error.

$$H_0 : F_{\tilde{r}_{DAX}} = F_{t_k} \quad vs. \quad H_1 : F_{\tilde{r}_{DAX}} \neq F_{t_k}.$$

Using the alternative notation we write

$$H_0 : \tilde{r}_{DAX} \sim t_k \quad vs. \quad H_1 : \tilde{r}_{DAX} \nsim t_k.$$

The number of degrees of freedom k are found via maximum likelihood estimation, (see Sect. 6.3.4 for estimation of copulae).

  The test statistic follows the Kolmogorov distribution, which was originally tabulated in Smirnov (1939). This distribution is independent of the assumed continuous univariate distribution under the null hypothesis.

```
> require(stats)
> dax      = EuStockMarkets[, 1]      # DAX index
> r.dax    = diff(log(dax))           # log-returns
> r.dax_st = scale(r.dax)             # standardisation
> l        = function(k, x){          # log-likelihood
+    -sum(dt(x, df = k, log = TRUE))
```

```
+ }
> k_ML = optimize(f = 1,            # optimize 1
+     interval = c(0, 30),          # range of k
+     x         = r.dax_st)$minimum # retrieve optimal k
> k_ML
[1] 11.56088
> ks.test(x = r.dax_st,             # test for t-dist.
+     y  = "pt",                     # t distribution function
+     df = k_ML)                     # estimated df


     One-sample Kolmogorov-Smirnov test

data:  r.dax_st
D = 0.063173, p-value = 7.194e-07
alternative hypothesis: two-sided
```

H_0 can be rejected for any significance level larger than the p-value. To test against other distributions, the parameter y should be set equal to the corresponding string variable of the cdf, like pnorm, pgamma, pcauchy etc.

To test whether the DAX and the FTSE log-returns follow the same distribution, one runs the following code in R.

```
> r.dax     = diff(log(EuStockMarkets[, 1]))
> ftse      = EuStockMarkets[, 4]         # FTSE index
> r.ftse    = diff(log(ftse))             # log-returns
> r.ftse_st = scale(r.ftse)               # standardisation
> ks.test(r.dax, r.ftse)                  # test with raw
                                          # log-returns

     Two-sample Kolmogorov--Smirnov test

data:  r.dax and r.ftse
D = 0.053792, p-value = 0.009223
alternative hypothesis: two-sided
> r.dax_st = scale(r.dax)                 # standardisation
> ks.test(r.dax_st, r.ftse_st)            # test with standardised
                                          # log-returns

     Two-sample Kolmogorov--Smirnov test

data:  r.dax_st and r.ftse_st
D = 0.034965, p-value = 0.2058
alternative hypothesis: two-sided
```

H_0 can be rejected for non-standardised log-returns, indicating that the DAX and the FTSE log-returns do not follow the same distribution. After standardisation of the log-returns, one can not reject H_0. Figure 5.11 illustrates these test results. The non-standardised log-returns have different means and standard deviations. Therefore the rejection of H_0 in the first test is due to different first and second moments. This example shows that the scaling of the variables can influence the test results.

The Kolmogorov-Smirnov test belongs to the group of exact tests, which are more reliable in smaller samples than asymptotic tests.

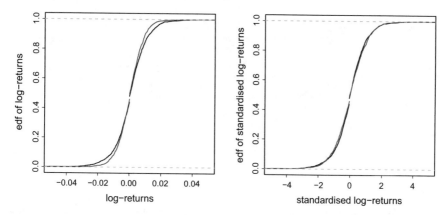

Fig. 5.11 Empirical cumulative distribution functions for DAX log-returns and FTSE log-returns.
Q BCS_EdfsDAXFTSE

Cramér–von Mises and Anderson–Darling test

Alternatives to the Kolmogorov-Smirnov test are the Cramér–von Mises and
Anderson–Darling tests. Instead of the largest distance in (5.9), they use the weighted
squared distance between the two distributions as the loss function:

$$\int_{-\infty}^{\infty} \{F(x) - G(x)\}^2 w(x) dG(x).$$

$F(x)$ is replaced by the ecdf of the sample and $G(x)$ is the cdf for the distribution
of X under H_0. The Anderson–Darling test uses different weights than does the
Cramér–von Mises test.

It is necessary to use ordered statistics to conduct both tests. Let $\{x_{(1)}, \ldots, x_{(n)}\}$ be
ordered random realizations of the rv X with $\mathsf{E}X = \mu$ and $\mathsf{Var}X = \sigma^2$. Furthermore,
one has to standardise these ordered realizations $z_{(i)} = \frac{x_{(i)} - \mu}{\sigma}$. In the following it is
assumed that $\mu = \bar{x}$ and $\sigma = s$.

For the Cramér–von Mises test, $w(x) = 1$ and the test statistic is

$$CM = \frac{1}{12n} + \sum_{i=1}^{n} \left\{ \frac{2i - 1}{2n} - G(z_{(i)}) \right\},$$

where n denotes the sample size. The distribution of the test statistic under H_0 can be
computed in R with pCvM and qCvM according to Csorgo and Farraway (1996). In the
following code, the Cramér–von Mises is used to test whether the standardised DAX
log returns follow the t-distribution with the same number of degrees of freedom as
found in previous subsection. In R the ordering of the realizations is automatically
done, but it is necessary to standardise them.

```
> require(goftest)
> r.dax_st = scale(diff(log(EuStockMarkets[, 1])))
> cvm.test(r.dax_st,              # Cramer von Mises test
+   null = "pt",                  # to test for t distr
+   df   = 11.56088)              # degrees of freedom

    Cramer-von Mises test of goodness-of-fit
    Null hypothesis: Student's t distribution
    with parameter df = 11.56088

data:  r.dax_st
omega2 = 3.0274, p-value = 6.457e-08
```

Again, the null hypothesis that $\tilde{r}_{DAX}$ comes from a rv that follows a $t_{11.56088}$-distribution can be rejected.

The Anderson–Darling test sets $w(x) = [G(x)\{1 - G(x)\}]^{-1}$, which leads to the test statistic

$$A^2 = -n - \frac{1}{n} \sum_{i=1}^{n} (2i - 1) \left[\log\{G(z_{(i)})\} + \log\{1 - G(z_{(n+1-i)})\} \right].$$

The distribution of the Anderson–Darling test statistic can be obtained by pAD and qAD. These functions are based on Marsaglia and Marsaglia (2004).

The following code uses the Anderson–Darling test to test whether the standardised DAX index log-returns follow a $t_{11.56088}$-distribution.

```
> require(goftest)
> r.dax_st = scale(diff(log(EuStockMarkets[, 1])))
> ad.test(r.dax_st, null = "pt", df = 11.56088)
                    # Anderson-Darling test
    Anderson-Darling test of goodness-of-fit
    Null hypothesis: Student's t distribution
    with parameter df = 11.56088

data:  r.dax_st
An = 17.715, p-value = 3.228e-07
```

The null hypothesis can be rejected for a significance level close to zero. Therefore the standardised log-returns of the DAX-index do not follow the $t_{11.56088}$-distribution. All three tests reject the null hypothesis of t-distributed log-returns of the DAX-index.

5.3.2 Tests for Normality

To verify whether a sample is generated by a normal distribution, the Shapiro–Wilk and the Jarque–Bera tests have been constructed. These tests are introduced in the following.

Shapiro–Wilk test

The Shapiro–Wilk test was developed in Shapiro and Wilk (1965). The specific form of H_0 leads to desirable efficiency properties. Especially in small samples, the power of this test is superior to other nonparametric tests, see Razali and Wah (2011).

The rv X is tested as to whether it follows a normal distribution.

$$H_0 : X \sim N \quad vs. \quad H_1 : X \nsim N.$$

The test statistic W is calculated by dividing the theoretically expected variance of a normal distribution by the realized sample variance:

$$W = \frac{\sigma^2}{(n-1)\tilde{s}^2} = \frac{(\sum_{i=1}^{n} c_i x_{(i)})^2}{\sum_{i=1}^{n} (x_{(i)} - \bar{x})^2},$$

where $\tilde{s}$ is the sample standard deviation and σ is the theoretically expected standard deviation, which is calculated from the ordered statistics $x_{(i)}$ as follows:

$$c = \frac{\tau^\top V^{-1}}{\sqrt{\tau^\top V^{-1} V^{-1} \tau}},$$

$$\sigma = \sum_{i=1}^{n} c_i x_{(i)}, \tag{5.10}$$

where τ is the vector of theoretically expected ordered values of x under normality and V the covariance matrix of τ.

The order statistics $x_{(i)}$ under H_0 follow the normal distribution and the probability of having an observation smaller than $x_{(i)}$ is $\Phi(x_{(i)})$. Values for τ are obtained under H_0 by

$$\tau_{(i)} = \Phi^{-1}\left(\frac{i - 3/8}{n + 1/4}\right).$$

Under H_0, the theoretical values $\tau_{(i)}$ for $x_{(i)}$ depend only on the sample size and the position i. Under H_0, the theoretical expected variance σ^2 should be close to the sample variance $\hat{\sigma}^2$. The test statistic W is bounded by

$$\frac{nc_1^2}{n-1} \leq W \leq 1.$$

If the test statistic W is close to one, H_0 can not be rejected. In this case the sample can be regarded as a realization of a normal rv. For low values of W it is likely that the null hypothesis is wrong and can be rejected. The distribution of the test statistic is tabulated in Shapiro and Wilk (1965).

The next example tests whether or not the DAX log-returns follow a normal distribution. The function `shapiro.test` is implemented in R as follows.

```
> r.dax = diff(log(EuStockMarkets[, 1]))
> shapiro.test(r.dax)        # by default H0: X ~ N(mu, sigma^2)

    Shapiro-Wilk normality test

data:   r.dax
W = 0.9538, p-value < 2.2e-16
```

The null hypothesis that r.dax follows a normal distribution can clearly be rejected.

```
> random = rnorm(1000) + 5 # for this sample H0 is not rejected
> shapiro.test(random)

    Shapiro-Wilk normality test

data:   random
W = 0.9987, p-value = 0.7089
```

Jarque–Bera test

An alternative test for normality is the Jarque–Bera test. The hypotheses are the same as for the Shapiro–Wilk test. The Jarque–Bera test considers the third and fourth moments of the distribution. Here $\hat{S}$ are the sample skewness and $\hat{K}$ the sample kurtosis, see Sect. 5.1.7. Then

$$JB = \frac{n}{6}\left(\hat{S}^2 + \frac{\hat{K}^2}{4}\right).$$

This test uses the results of Chap. 4 for the moments of the normal distribution for which skewness and excess kurtosis should be both zero. Two parameters are estimated to compute the test statistic, therefore the statistic follows the χ_2^2 distribution.

There is an implementation for the Jarque–Bera test in R, which requires the package tseries.

```
> require(tseries)
> r.dax = diff(log(EuStockMarkets[, 1]))
> jarque.bera.test(r.dax) # by default H0: X ~ N(mu, sigma^2)

Jarque-Bera Test

data:   r.dax
X-squared = 3149.641, df = 2, p-value < 2.2e-16
```

The p-values provided by R for daily DAX log-returns are identically small for the two tests, but this is not true for the truly normal sample. In general, inference with both tests might lead to different conclusions.

Most of these tests are also provided by the package fBasics and functions ksnormTest, shapiroTest, jarqueberaTest, jbTest, adTest and cvmTest.

5.3.3 Wilcoxon Signed Rank Test and Mann–Whitney U Test

The Kolmogorov–Smirnov test assumes an interval or ratio scale for the variable of interest. Wilcoxon (1945) developed two tests that also work for ordinal data: the Wilcoxon signed rank and rank sum tests. The latter is also known as the Mann–Whitney U test.

The Wilcoxon signed rank test is an asymptotic test for the median $\tilde{x}_{0.5}$ of the sample $\{x_1, \ldots, x_n\}$, see Sect. 5.1.5.

$$H_0 : \tilde{x}_{0.5} = a \quad vs. \quad H_1 : \tilde{x}_{0.5} \neq a,$$

where a is an assumed value. For two samples $\{x_{1,1}, \ldots, x_{1,n_1}\}$ and $\{x_{2,1}, \ldots, x_{2,n_2}\}$ with sample sizes n_1 and n_2, the hypotheses are

$$H_0 : \tilde{x}_{1,0.5} = \tilde{x}_{2,0.5} \quad vs. \quad H_1 : \tilde{x}_{1,0.5} \neq \tilde{x}_{2,0.5}.$$

The algorithm of the Wilcoxon signed rank test for two samples can be written as follows:

1. Randomly draw $n_s = \min(n_1, n_2)$ observations from the larger sample;
2. Calculate $s_i = \text{sign}(x_{1,i} - x_{2,i})$ and $d_i = |x_{1,i} - x_{2,i}|$ for the paired samples;
3. Compute the ranks R_i of d_i ascending from 1;
4. Then the test statistic is $W = |\sum_{i=1}^{n_s} s_i R_i|$.

The test statistic has the asymptotic distribution

$$W \xrightarrow[n_s \to \infty]{\mathcal{L}} N(0.5, \sigma_W^2), \tag{5.11}$$

$$\text{with } \sigma_W = \sqrt{\frac{n_s(n_s+1)(2n_s+1)}{6}}.$$

Thus the Wilcoxon signed rank test checks whether two samples come from the same population, in which case the mean of the weighted $sign()$ operator is 0.5, just as for a fair coin toss. If the statistic is close to 0.5, positive and negative differences are equally likely. The second set can also be a vector $1_n a$ if one wants to test against a specific constant a.

Note the test statistic follows the normal distribution only asymptotically. However, ranks are ordinal and not metric, therefore the assumption of a normal distribution is not appropriate in finite and small samples. It is necessary to correct the test statistic for continuity, which is done by default in R.

Consider as an example the popularity of American presidents in the past. For this we use the dataset `presidents` and denote the sample by $\{x_1, \ldots, x_n\}$, which contains quarterly approval ratings in percentages for the President of the United States from the first quarter of 1945 to the last quarter of 1974. To verify that the median ranking is at least 50, the following hypotheses are tested:

$$H_0 : \tilde{x}_{0.5} \leq a \quad vs. \quad H_1 : \tilde{x}_{0.5} > a,$$

Here an artificial sample $y = 1_n a = 50, \ldots, 50$ with the same number of elements as the number of observations n for the dataset `presidents` is created. Therefore the test in R compares the actual dataset with an artificial dataset with all elements equal to c. It is necessary that the samples for the Wilcoxon signed rank test are paired in R, which means they have the same number of elements. If H_0 is rejected, the presidents have a simple majority behind them.

```
> s = 50                          # maximum median under H0
> y = rep(s, length(presidents))  # vector of constants
> wilcox.test(presidents, y,      # test for presidents
+    alternative = "greater",     # specifies H1
+    paired      = TRUE)          # signed rank test

    Wilcoxon signed rank test with continuity correction

data:  presidents and y
V = 4613.5, p-value = 3.298e-05
alternative hypothesis: true location shift is greater than 0
```

The p-value turns out to be close to zero and we can reject the hypothesis that the true value of the approval ratings is at most 50.

Unlike the signed rank test, the Mann–Whitney U test can also be used for non-paired data. Let F_1 and F_2 denote the distributions of two variables

$$H_0 : F_1(x) = F_2(x) \quad vs. \quad H_1 : F_1(x) = F_2(x - a) \quad \forall x \in \mathbb{R}$$

The Wilcoxon rank sum test is then calculated as follows:

1. Merge two samples into one sample and rank them;
2. Let R_j be the sum of all ranks in the combined sample for observations of sample $j \in \{1, 2\}$;
3. $U_j = n_1 n_2 + \frac{1}{2} n_j (n_j + 1) - R_j$, n_j is the size of sample $j \in \{1, 2\}$, $R_1 + R_2 = \frac{1}{2} n(n + 1)$, where $n = n_1 + n_2$;
4. $U = \min(U_1, U_2)$;
5. $U \xrightarrow[n \to \infty]{\mathcal{L}} N\left\{ n_1 n_2 / 2, \sqrt{n_1 n_2 (n_1 + n_2 + 1)/12} \right\}$.

The core idea of the test is that half the maximal possible sum of ranks is deducted from the actual sum of ranks. If both samples are from the same distribution, this statistic should be close to $\frac{n_1 n_2}{2}$.

Now one may ask the question whether President Nixon's popularity was significantly lower than that of his predecessors. The dataset is split into two parts: one containing the realizations from the previous presidents and another set for President Nixon:

```
> Other = presidents[1:96]        # sample for other presidents
> Nixon = presidents[97:118]      # sample for Nixon
> wilcox.test(Nixon, Other)       # test for popularity

    Wilcoxon rank sum test with continuity correction

data:  Nixon and Other
W = 647.5, p-value = 0.03868
alternative hypothesis: true location shift is not equal to 0
```

The hypothesis that the medians are equal can clearly be rejected, because the obtained p-value is smaller than 5%. The difference between the sample medians is too big for the distributions to be considered equal. Nixon, with a median approval rating of 49%, was significantly less popular than other presidents, with 61%.

5.3.4 Kruskal–Wallis Test

The tests discussed above considered two samples. They can not be used to check for the equality of more than two samples. Consider a test that rejects the null hypothesis of pairwise equal distributions for three variables X, Y and Z if at least for one pair a two-sample test rejects the equality of distributions at the significance level α. It would be wrong to assume that the joint significance of this test is $\alpha = 0.05$ because the probability for the test to favour equality between X and Y, Y and Z and X and Z is in fact the probability of not rejecting three times, which is equal to $1 - (1 - 0.05)^3 = 1 - 0.86 = 0.14$ which means an α of 0.14 and not 0.05.

Kruskal and Wallis (1952) developed an extension of the Mann–Whitney U test that solves this problem. The null hypothesis is rejected if at least one sample distribution has a different mean than the other distributions. Let $l = \{1, \ldots, m\}$ be an index of the considered samples and a be a constant,

$$H_0 : F_1(x) = \ldots = F_m(x) \quad vs. \quad H_1 : F_1(x) = \ldots = F_l(x+a) = \ldots = F_m(x), \quad \forall x \in \mathbb{R}.$$

The test statistic is defined as

$$K = (n - 1)\frac{\sum_{l=1}^{m} n_l(\bar{R}_l - \bar{R})^2}{\sum_{l=1}^{m}\sum_{j=1}^{n_l}(R_{jl} - \bar{R})^2},$$

where $n = \sum_{l=1}^{m} n_l$ and $\bar{R}_l = \sum_{j=1}^{n_l} R_{jl}/n_l$ denotes the average of all ranks allocated within sample l. $\bar{R} = \sum_{j=1}^{n_l}\sum_{l=1}^{m} R_{jl}/n$ is the overall average of all ranks in all samples and R_{jl} is the rank for the pooled sample of observation j in sample l. The test statistic follows the χ_k^2 distribution, where $k = m - 1$. In the following example we compare again the popularity of the ruling president for every single decade from the first quarter of 1945 to the last quarter of 1974. We define a variable

of starting points for each group. A Kruskal–Wallis test for the null hypothesis that
the popularity did not change significantly is executed by

```
> decades = c(rep(1, length.out = 20), # group indicator for decades
+    rep(2:3, each = 40),
+    rep(4,   length.out = 20))
> kruskal.test(presidents, decades)    # Kruskal-Wallis test

    Kruskal--Wallis rank sum test

data:  presidents and decades
Kruskal-Wallis chi-squared = 12.2607, df = 3, p-value = 0.006541
```

Over the decades, the popularity of the presidents varies significantly. This means
that the null hypothesis of equal locations of the distributions $\bar{R} = \bar{R}_l$, for all l can
be rejected at a significance level close to zero.

Chapter 6
Multivariate Distributions

Though this be madness, yet there is method in't.

— William Shakespeare, Hamlet

The preceding chapters discussed the behaviour of a single rv. This chapter introduces the basic tools of statistics and probability theory for multivariate analysis, where the relations between d rvs are considered. At first we present the basic tools of probability theory used to describe a multivariate rv, including the marginal and conditional distributions and the concept of independence.

The normal distribution plays a central role in statistics because it can be viewed as an approximation and limit of many other distributions. The basic justification for this relies on the central limit theorem. This is done in the framework of sampling theory, together with the main properties of the multinormal distribution.

However, a multinormal approximation can be misleading for data which is not symmetric or has heavy tails. The need for a more flexible dependence structure and arbitrary marginal distributions has led to the wide use of copulae for modelling and estimating multivariate distributions.

6.1 The Distribution Function and the Density Function of a Random Vector

For $X = (X_1, X_2, \ldots, X_d)^\top$ a random vector, the cdf is defined as

$$F(x) = P(X \leq x) = P(X_1 \leq x_1, \ldots, X_d \leq x_d).$$

This function describes the joint behaviour of components of the vector. If X is discrete, then there exists a joint density function $p(\cdot)$ given by

© Springer International Publishing AG 2017
W.K. Härdle et al., *Basic Elements of Computational Statistics*,
Statistics and Computing, DOI 10.1007/978-3-319-55336-8_6

$$p(x) = P(X = x) = P(X_1 = x_1, \ldots, X_d = x_d).$$

Assume that $X_1, \ldots, X_d$ are continuous rvs satisfying

$$\frac{\partial^d}{\partial x_1 \ldots \partial x_d} F(x) = \frac{\partial^d}{\partial x_{i_1} \ldots \partial x_{i_d}} F(x),$$

for all permutations $i_1, \ldots, i_d$ of $1, \ldots, d$. Then the joint density function is given by

$$f(x) = \frac{\partial^d F(x)}{\partial x} = \frac{\partial^d F(x_1, \ldots, x_d)}{\partial x_1 \cdots \partial x_d}.$$

Note that, as in the one-dimensional case,

$$\int_{-\infty}^{\infty} \ldots \int_{-\infty}^{\infty} f(u) \, du_1 \ldots du_d = 1.$$

If the density function is differentiable, then

$$F(x) = \int_{-\infty}^{x_1} \ldots \int_{-\infty}^{x_d} f(u_1, \ldots, u_d) du_d \ldots du_1.$$

If we partition $(X_1, \ldots, X_d)^\top$ as $X_k^* = (X_{i_1}, \ldots, X_{i_k})^\top \in \mathbb{R}^k$ and $X_{-k}^* = (X_{i_{k+1}}, \ldots, X_{i_d})^\top \in \mathbb{R}^{d-k}$, then the function defined by

$$F_{X_k^*}(x_{i_1}, \ldots, x_{i_k}) = P(X_{i_1} \leq x_{i_1}, \ldots, X_{i_k} \leq x_{i_k})$$

is called the *k-dimensional marginal cdf* and is equal to F evaluated at $(x_{i_1}, \ldots, x_{i_k})$ and x_{-k}^* set to infinity. For continuous variables, the marginal pdf can be computed from the joint density by "integrating out" irrelevant variables

$$f_{X_k}(x_{i_1}, \ldots, x_{i_k}) = \int_{-\infty}^{\infty} \ldots \int_{-\infty}^{\infty} f(x_1, \ldots, x_d) dx_{i_{k+1}} \ldots dx_{i_d}.$$

For discrete X, the marginal probability is given by

$$p_{X_k}(x_{i_1}, \ldots, x_{i_k}) = \sum_{x_{i_{k+1}}, \ldots, x_{i_d}} p(x_1, \ldots, x_d).$$

Generally speaking, the following theory works in any dimension $d \geq 2$ but in some cases, for simplicity, we restrict ourselves to the two-dimensional case $X = (X_1, X_2)^\top$. Let us consider the conditional pdf of X_2 given $X_1 = x_1$

$$f(x_2 \mid x_1) = \frac{f(x_1, x_2)}{f_{X_1}(x_1)}.$$

Two rvs X_1 and X_2 are said to be independent if $f(x_1, x_2) = f_{X_1}(x_1) f_{X_2}(x_2)$, which implies $f(x_1 \mid x_2) = f_{X_1}(x_1)$ and $f(x_2 \mid x_1) = f_{X_2}(x_2)$. Independence can be interpreted as follows: knowing $X_2 = x_2$ does not change the probability assessments on X_1, and vice versa.

In multivariate statistics, we observe the values of a multivariate rv X and obtain a sample $\mathcal{X} = \{x_i\}_{i=1}^n$. Under random sampling, these observations are considered to be realisations of a sequence of i.i.d. rvs $X_1, \ldots, X_n$, where each X_i has the same distribution as the *parent* or *population* rv X.

$$\mathcal{X} = \{x_i\}_{i=1}^n = \begin{pmatrix} x_{11} & \cdots & x_{1d} \\ \vdots & \ddots & \vdots \\ x_{n1} & \cdots & x_{nd} \end{pmatrix},$$

where x_{ij} is the ith realisation of the jth element of the random vector X. The idea of statistical inference for a given random sample is to analyse the properties of the population variable X. This is typically done by analysing some characteristics of its distribution.

6.1.1   Moments

Expectation

The first-order moment of X, often called the *expectation*, is given by

$$\mathsf{E}\,X = \begin{pmatrix} \mathsf{E}\,X_1 \\ \vdots \\ \mathsf{E}\,X_d \end{pmatrix} = \int x f(x) dx = \begin{pmatrix} \int x_1 f(x) dx \\ \vdots \\ \int x_d f(x) dx \end{pmatrix} = \begin{pmatrix} \mu_1 \\ \vdots \\ \mu_d \end{pmatrix} = \mu.$$

Accordingly, the expectation of a matrix of random elements has to be understood component by component, see Definition 4.3. The operation of forming expectations is linear

$$\mathsf{E}\,(\alpha X + \beta Y) = \alpha \mathsf{E}\,X + \beta \mathsf{E}\,Y, \tag{6.1}$$

with $\alpha, \beta \in \mathbb{R}$ and $X = (X_1, \ldots, X_d)^\top$ and $Y = (Y_1, \ldots, Y_d)^\top$ being random vectors of the same dimension. If $\mathcal{A}(q \times d)$ is a matrix of real numbers, we have

$$\mathsf{E}(\mathcal{A}X) = \mathcal{A}\,\mathsf{E}\,X.$$

When X and Y are independent,

$$\mathsf{E}(XY^\top) = \mathsf{E}\,X\,\mathsf{E}\,Y^\top.$$

Statistics describing the *center of gravity* of the n observations in $\mathbb{R}^d$ are given by the vector $\bar{x}$ of the mean values $\bar{x}_j$ (see Sect. 5.1.5) for $j = (1, \ldots, d)$ and can be obtained by

$$\bar{x} = \begin{pmatrix} \bar{x}_1 \\ \vdots \\ \bar{x}_d \end{pmatrix} = \frac{1}{n}\mathcal{X}^\top 1_n.$$

It is easy to show that $\bar{x}$ is an unbiased estimator of the expectation $\mathsf{E}\,X$.

R offers several functions with which to calculate the sample mean. The function mean is applicable to vector, matrix and data.frame types. If the sample is represented by a matrix A, mean(A) returns the average over all elements in the matrix. Below we applied this function to the data set women, which contain the average heights and weights for American women aged 30–39.

```
> women.m = as.matrix(women)  # convert data.frame to matrix
> mean(women.m)               # mean of matrix
[1] 100.8667
```

rowMeans and colMeans calculate the averages by rows and columns of the matrix, or data frame respectively.

```
> rowMeans(women.m)              # averages by rows
 [1]   86.5   88.0   90.0   92.0   94.0   96.0   98.0
 [8]  100.0  102.5  104.5  107.0  109.5  112.0  115.0  118.0
> colMeans(women.m)              # averages by columns
   height     weight
  65.0000   136.7333
```

Covariance matrix

The matrix

$$\mathsf{Var}\,X = \Sigma = \mathsf{E}(X - \mu)(X - \mu)^\top = \mathsf{E}(XX^\top) - \mu\mu^\top$$

is the *(theoretical) covariance matrix*, also called the centred second moment. It is positive semi-definite, i.e. $\Sigma \geq 0$, with elements $\Sigma = (\sigma_{X_i X_j})$. The off-diagonal elements are $\sigma_{X_i X_j} = \mathsf{Cov}(X_i, X_j)$ and the diagonal elements are $\sigma_{X_i X_i} = \mathsf{Var}(X_i)$, $i, j = 1, \ldots, d$, where

$$\mathsf{Cov}(X_i, X_j) = \mathsf{E}(X_i X_j) - \mu_i \mu_j,$$
$$\mathsf{Var}\,X_i = \mathsf{E}\,X_i^2 - \mu_i^2.$$

Writing $X \sim (\mu, \Sigma)$ means that X is a random vector with mean vector μ and covariance matrix Σ. The variance of the linear transformation of the variables satisfies

$$\text{Var}(\mathcal{A}X) = \mathcal{A}\text{Var}(X)\mathcal{A}^\top = \sum_{i,j} a_i a_j \sigma_{X_i X_j},$$

$$\text{Var}(\mathcal{A}X + b) = \mathcal{A}\,\text{Var}(X)\mathcal{A}^\top, \tag{6.2}$$

where $\mathcal{A}$ is the $(q \times d)$ matrix an b a q-dimensional vector. For $X \in \mathbb{R}^d$ and $Y \in \mathbb{R}^q$, the $(d \times q)$ covariance matrix of $X \sim (\mu_X, \Sigma_{XX})$ and $Y \sim (\mu_Y, \Sigma_{YY})$ is

$$\Sigma_{XY} = \text{Cov}(X, Y) = \text{E}(X - \mu_X)(Y - \mu_Y)^\top.$$

This result is obtained from

$$\text{Cov}(X, Y) = \text{E}(XY^\top) - \mu_X \mu_Y^\top = \text{E}(XY^\top) - \text{E}\,X\,\text{E}\,Y^\top.$$

It follows that if X and Y are independent, then $\text{Cov}(X, Y) = 0$. $\text{E}(XX^\top)$ provides the second non-central moment of X

$$\text{E}(XX^\top) = \{\text{E}(X_i X_j)\}, \text{ for } i, j = 1, \ldots, d.$$

Linear functions of random vectors have the following covariance properties:

$$\text{Cov}(X + Y, Z) = \text{Cov}(X, Z) + \text{Cov}(Y, Z);$$
$$\text{Var}(X + Y) = \text{Var}(X) + \text{Cov}(X, Y) + \text{Cov}(Y, X) + \text{Var}(Y);$$
$$\text{Cov}(\mathcal{A}X, \mathcal{B}Y) = \mathcal{A}\,\text{Cov}(X, Y)\mathcal{B}^\top.$$

The sample covariance matrix is used to estimate the second-order moment

$$\tilde{\Sigma} = n^{-1}\mathcal{X}^\top\mathcal{X} - \overline{x}\,\overline{x}^\top. \tag{6.3}$$

For small sample sizes, $\tilde{\Sigma}$ is biased. An unbiased estimator of the second moment is

$$\hat{\Sigma} = \frac{1}{n-1}\mathcal{X}^\top\mathcal{X} - \frac{n}{n-1}\overline{x}\,\overline{x}^\top. \tag{6.4}$$

Equation (6.4) can be written equivalently in scalar form or based on the centering matrix $\mathcal{H} = \mathcal{I}_n - n^{-1}1_n 1_n^\top$

$$\hat{\Sigma} = (n-1)^{-1}(\mathcal{X}^\top\mathcal{X} - n^{-1}\mathcal{X}^\top 1_n 1_n^\top \mathcal{X})$$
$$= (n-1)^{-1}\mathcal{X}^\top\mathcal{H}\mathcal{X}. \tag{6.5}$$

These formulas are implemented directly in R. The function cov returns the empirical covariance matrix of the given sample matrix $\mathcal{X}$. Its argument could be of type `data.frame`, `matrix`, or consist of two `vectors` of the same size. The following code presents possible calculations of $\hat{\Sigma}$:

```
> women.m = as.matrix(women)
> n       = dim(women.m)[1]; n
[1] 15
> meanw   = colMeans(women)
> cov1    =                                        # using (6.4)
+    (t(women.m) %*% women.m - n * meanw %*% t(meanw)) / (n - 1)
> H       = diag(1, n) - 1 / n * rep(1, n) %*% t(rep(1, n))
> cov2    = t(women.m) %*% H %*% women.m / (n - 1)  # using (6.5)
> cov3    = cov(women)                              # for data.frame
> cov4    = cov(women.m)                            # for matrix
```

As expected, all the matrices `cov1`, `cov2`, `cov3` and `cov4` return the same result.

```
          height   weight
height       20   69.0000
weight       69  240.2095
```

The internal function `cov` is twice as fast as manual methods with or without a predetermined centred matrix, and independent of sample size.

If the arguments of `cov` are two vectors x and y, then the result is their covariance.

```
> cov(women$height, women$weight)
[1] 69
```

The correlation ρ_{X_i,X_j} between two rvs X_i and X_j is given by

$$\rho_{X_i,X_j} = \frac{\mathsf{Cov}(X_i, X_j)}{\sqrt{\mathsf{Var}\, X_i\, \mathsf{Var}\, X_j}} = \frac{\sigma_{X_i,X_j}}{\sqrt{\sigma_{X_i}\sigma_{X_j}}}.$$

Thus, the correlation matrix is $\mathcal{P}_{\mathcal{X}} = \{\rho_{X_i,X_j}\}$. Similar to the covariance matrix, for rvs X and Y

$$\mathsf{Cor}(X, Y) = (\mathsf{Var}\, X)^{-1/2}\, \mathsf{Cov}(X, Y)(\mathsf{Var}\, Y)^{-1/2}.$$

The unbiased sample correlation is given by

$$\hat{\rho}_{X_i,X_j} = \frac{n\sum_{m=1}^{n} x_{im}x_{jm} - \sum_{m=1}^{n} x_{im} \sum_{m=1}^{n} x_{jm}}{\sqrt{n\sum_{m=1}^{n} x_{im}^2 - (\sum_{m=1}^{n} x_{im})^2}\sqrt{n\sum_{m=1}^{n} x_{jm}^2 - (\sum_{m=1}^{n} x_{jm})^2}}.$$

The calculation of the sample correlation in R is done by the atomic function `cor`, which is similar to `cov`. The covariance matrix $\hat{\Sigma}$ may be converted into a correlation matrix using the function `cov2cor`.

```
> cor(women);
> cov2cor(cov(women))
          height      weight
height 1.0000000   0.9954948
weight 0.9954948   1.0000000
```

The linear correlation is sensitive to outliers, and is invariant only under strictly increasing linear transformations. Alternative rank correlation coefficients, which are less sensitive to outliers, are Kendall's τ and Spearman's ρ_S. If F is a continuous bivariate cumulative distribution function and (X_1, X_2), (X'_1, X'_2) are independent random pairs with the same distribution F, then Kendall's τ is

$$\tau = P\{(X_1 - X'_1)(X_2 - X'_2) > 0\} - P\{(X_1 - X'_1)(X_2 - X'_2) < 0\}.$$

Assuming that the marginal distributions of X_1 and X_2 are given by F_1 and F_2, respectively. Spearman's ρ_S is defined as

$$\rho_S = \frac{\text{Cov}\{F_1(X_1), F_2(X_2)\}}{\sqrt{\text{Var}\{F_1(X_1)\}\,\text{Var}\{F_2(X_2)\}}}.$$

Both rank-based correlation coefficients are invariant under strictly increasing transformations and measure the 'average dependence' between X_1 and X_2. The empirical $\hat{\tau}$ and $\hat{\rho}_S$ are calculated by

$$\hat{\tau} = \frac{4}{n(n-1)} P_n - 1, \tag{6.6}$$

$$\hat{\rho}_S = \frac{\sum_{i=1}^{n}(R_i - \overline{R})^2 (S_i - \overline{S})^2}{\sqrt{\sum_{i=1}^{n}(R_i - \overline{R})^2 \sum_{i=1}^{n}(S_i - \overline{S})^2}}, \tag{6.7}$$

where P_n is the number of concordant pairs, i.e., the number of pairs (x_{1k}, x_{2k}) and (x_{1m}, x_{2m}) of points in the sample for which

$$x_{1k} < x_{1m} \quad \text{and} \quad x_{2k} < x_{2m}$$

$$\text{or}$$

$$x_{1k} > x_{1m} \quad \text{and} \quad x_{2k} > x_{2m}.$$

The R_i and S_i in (6.7) represent the position of the observation in a sorted by size list of all observations (statistical rank). These two correlation coefficients are implemented by the function cor, using the parameter method. In the following listing we applied cor function to the dataset cars, which contain the speed of cars and the distances taken to stop (data were recorded in the 1920s).

```
> cor(cars$speed, cars$dist)
[1] 0.8068949
> cor(cars$speed, cars$dist, method = "kendall")
[1] 0.6689901
> cor(cars$speed, cars$dist, method = "spearman")
[1] 0.8303568
```

The robustness to outliers and invariance under monotone increasing transformations are illustrated in Fig. 6.1. The outlier in the left panel of the figure makes the linear Pearson correlation coefficient $\rho = 0.66$, while Spearman's ρ_S and Kendall's

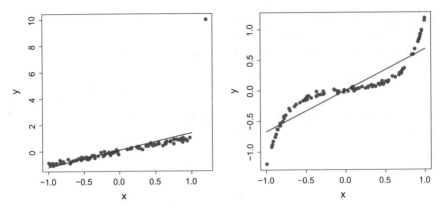

Fig. 6.1 Linear fit for the linearly correlated data with an outlier (*left*) and almost perfectly dependent monotone transformed data (*right*). Q BCS_CopulaInvarOutlier

τ are equal to 0.98 and 0.88, respectively. The same is observed on the right side of Fig. 6.1: the nonlinear but monotone transformation of the almost perfectly dependent data results in $\rho = 0.892$ but $\rho_S = 0.996$ and $\tau = 0.956$.

6.2   The Multinormal Distribution

The multivariate normal distribution is one of the most widely used multivariate distributions. A random vector X is said to be normally distributed with mean μ and covariance $\Sigma > 0$, or $X \sim N_d(\mu, \Sigma)$, if it has the following density function:

$$f(x) = |2\pi\Sigma|^{-1/2} \exp\left\{-\frac{1}{2}(x - \mu)^\top \Sigma^{-1}(x - \mu)\right\}. \tag{6.8}$$

As with the univariate normal distribution, the multinormal distribution does not have an explicit form for its cdf, i.e. $\Phi(x) = \int_{-\infty}^{x_1} \cdots \int_{-\infty}^{x_d} f(u)du$.

The multivariate t-distribution is closely related to the multinormal distribution. If $Z \sim N_d(0, \mathcal{I}_d)$ and $Y^2 \sim \chi_m^2$ are independent rvs, a t-distributed rv T with m degrees of freedom can be defined by

$$T = \sqrt{m}\frac{Z}{Y}.$$

Moreover, the multivariate t-distribution belongs to the family of d-dimensional spherical distributions, see Fang and Zhang (1990).

R offers several independent packages for the multinormal and multivariate t distributions, namely fMultivar by Wuertz et al. (2009b), mvtnorm by Genz and Bretz (2009) and Genz et al. (2012), and mnormt by Genz and Azzalini (2012).

The package `fMultivar` was specifically developed for bivariate distributions, including the multinormal, the t distribution, and the Cauchy distribution. It also allows of non-parametric density estimation, see Sect. 5.1.4. The packages `mvtnorm` and `mnormt` are for multinormal and multivariate t distributions of dimensions $d \geq 2$. All three packages contain methods for evaluating the density and distribution functions at a given point and for a given mean and covariance or correlation matrix. The calculation of the density is unproblematic because of its explicit form and is easily done by the following listing with $\Sigma(\sigma_{ij})$, $\sigma_{12} = \sigma_{21} = 0.7$, $\sigma_{11} = \sigma_{22} = 1$, $\mu = (0, 0)^\top$ and the point at which the density is calculated $x = (0.3, 0.4)^\top$.

```
> require(mvtnorm)
> lsigma = matrix(c(1, 0.7, 0.7, 1), ncol = 2)
> lsigma                                        # covariance matrix
      [,1] [,2]
[1,]   1.0  0.7
[2,]   0.7  1.0
> lmu       = c(0, 0)                            # mean vector
> x         = c(0.3, 0.4)                        # value at which to eval.
> dmvnorm(x, mean = lmu, sigma = lsigma)         # density at point x
[1] 0.2056464
```

From (6.8) and Fig. 6.2, one sees that the density of the multinormal distribution is constant on ellipsoids of the form

$$(x - \mu)^\top \Sigma^{-1} (x - \mu) = a^2. \tag{6.9}$$

The half-lengths of the axes in the contour ellipsoid are $\sqrt{a^2 \lambda_i}$, where λ_i are the eigenvalues of Σ. If Σ is a diagonal matrix, the rectangle circumscribing the contour ellipse has sides of length $2a\sigma_i$ and is thus naturally proportional to the standard deviations of X_i ($i = 1, 2$).

The distribution of the quadratic form in (6.9) is given in the next theorem.

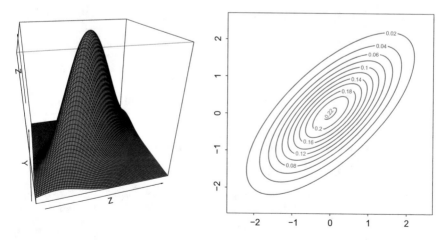

Fig. 6.2 Density of N_2, with $\rho_{12} = 0.7$. ⚙ BCS_BinormalDensity

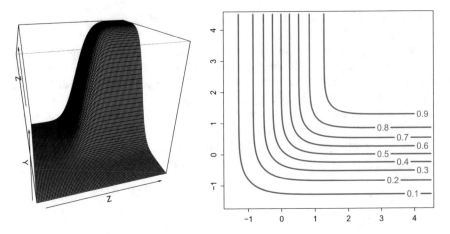

Fig. 6.3 cdf of N_2, with $\rho_{12} = 0.7$. Q BCS_NormalProbability

Theorem 6.1 *If $X \sim N_d(\mu, \Sigma)$, then rv $\mathcal{U} = (X - \mu)^\top \Sigma^{-1}(X - \mu)$ has a χ_d^2 distribution.*

As mentioned above, calculating the cdf is more problematic. The method `pmvnorm` from the `mvtnorm` package implements two algorithms to evaluate the normal distribution. The first is based on the randomised Quasi Monte Carlo procedure by Genz (1992) and (1993), which is applicable to singular and non-singular covariance structures of dimension $d \leq 1000$. The second algorithm by Miwa et al. (2003) is only applicable for small dimensions, $d \leq 20$ and non-singular covariance matrices. The package `mnormt` uses only the latter method. As a result of the evaluation of the distribution function, `pmvnorm` also returns the estimated absolute error (Fig. 6.3).

```
> lsigma    = matrix(c(1,  0.7,  0.7,  1),  ncol = 2)
> lmu       = c(0,  0)
> x         = c(0.3,  0.4)
> pmvnorm(x, mean = lmu, sigma = lsigma)
[1] 0.2437731
> attr(, "error")
[1] 1e-15
> attr(, "msg")
[1] "Normal Completion"
```

Simulation techniques, see Chap. 9, are implemented in several packages. The following code demonstrates the simplest case, $d = 2$ using `mvtnorm`.

```
> lsigma    = matrix(c(1,  0.7,  0.7,  1),  ncol = 2)
> lmu       = c(0,  0)
> # set.seed(2^11 - 1)                          # set the seed, see Chapter 9
> rmvnorm(5, mean = lmu, sigma = lsigma)        # sample 5 observations
             [,1]        [,2]
[1,] -0.1508041 -0.7374920
[2,]  1.0681719  0.3484549
[3,] -0.3611665 -0.3819258
[4,]  0.8141042  0.5995360
[5,]  1.4857324  1.4820885
```

Table 6.1 Multinormal distribution in R

	fMultivar ($d = 2$)	mvtnorm ($d \geq 2$)	mnormt ($d \geq 2$)
cdf (probability)	pnorm2d	pmvnorm (GenzBretz, Miwa, TVPACK)	pmnorm
pdf (density)	dnorm2d	dmvnorm	dmnorm
simulation	rnorm2d	rmvnorm	rmnorm
quantiles	n.a.	qmvnorm	n.a.

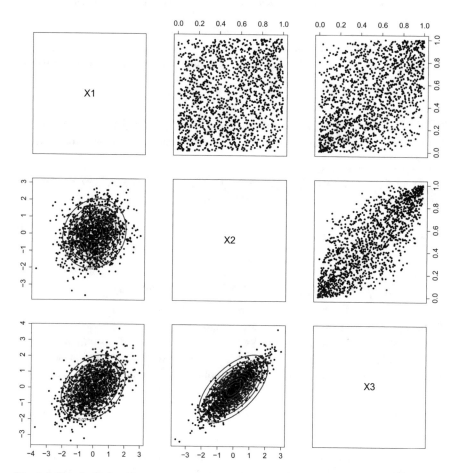

Fig. 6.4 Sample $\mathcal{X}$ from N$_3$, with $\rho_{12} = 0.2$, $\rho_{13} = 0.5$, $\rho_{23} = 0.8$ and $n = 1500$. Plots for estimated marginal distributions of $X_{1,2,3}$ are in the *upper triangle* and contour ellipsoids for the bivariate normal densities in the *lower triangle* of the matrix. **Q** BCS_NormalCopulaContour

The package mvtnorm is the only package with a method of calculating the quantiles of the multinormal distribution, namely qmvtnorm. Table 6.1 lists and compares the methods from all these packages (Fig. 6.4). If $\mu = 0$ and $\Sigma = \mathcal{I}_d$, then

X is said to be standard normally distributed. Using linear transformations, one can shift and scale the standard normal vector to form non-standard normal vectors.

Theorem 6.2 (Mahalanobis transformation) *Let $X \sim N_d(\mu, \Sigma)$ and $Y = \Sigma^{-1/2} (X - \mu)$. Then*

$$Y \sim N_d(0, \mathcal{I}_d),$$

i.e., the elements $Y_j \in \mathbb{R}$ are independent, one-dimensional $N(0, 1)$ variables.

Note that the Mahalanobis transformation in Eq. (6.2) yields an rv $Y = (Y_1, \ldots, Y_d)^\top$ composed of independent one-dimensional $Y_j \sim N_1(0, 1)$. We can create $N_d(\mu, \Sigma)$ variables on the basis of $N_d(0, \mathcal{I}_d)$ variables using the inverse linear transformation

$$X = \Sigma^{1/2}Y + \mu.$$

Using (6.1) and (6.2), we can verify that $E(X) = \mu$ and $Var(X) = \Sigma$. The following theorem is useful because it presents the distribution of a linearly transformed variable.

Theorem 6.3 *Let $X \sim N_d(\mu, \Sigma)$, $\mathcal{A}$ $(p \times p)$ and $b \in \mathbb{R}^p$, where $\mathcal{A}$ is non-singular. Then $Y = \mathcal{A}X + b$ is again a p-variate normal, i.e.*

$$Y \sim N_p(\mathcal{A}\mu + b, \mathcal{A}\Sigma\mathcal{A}^\top). \tag{6.10}$$

6.2.1 Sampling Distributions and Limit Theorems

Statistical inference often requires more than just the mean or the variance of a statistic. We need the sampling distribution of a statistic to derive confidence intervals and define rejection regions in hypothesis testing for a given significance level. Theorem 6.4 gives the distribution of the sample mean for a multinormal population.

Theorem 6.4 *Let $X_1, \ldots, X_n$ be i.i.d. with $X_i \sim N_d(\mu, \Sigma)$. Then $\overline{X} \sim N_d(\mu, n^{-1}\Sigma)$.*

In multivariate statistics, the sampling distributions are often more difficult to derive than in the previous theorem. They might be so complicated in closed form that limit theorem approximations must be used. The approximations are only valid when the sample size is large enough. In spite of this restriction, these approximations make complicated situations simpler. The central limit theorem shows that even if the parent distribution is not normal, the mean $\overline{X}$ has, for large sample sizes, an approximately normal distribution.

Theorem 6.5 (Central Limit Theorem (CLT)) *Let $X_1, \ldots, X_n$ be i.i.d. with $X_i \sim (\mu, \Sigma)$. Then the distribution of $\sqrt{n}(\overline{X} - \mu)$ is asymptotically $N_d(0, \Sigma)$, i.e.,*

$$\sqrt{n}(\overline{X} - \mu) \xrightarrow{\mathcal{L}} N_d(0, \Sigma) \quad \text{as } n \to \infty.$$

This asymptotic normality of the distribution is often used to construct confidence intervals for the unknown parameters. If the covariance matrix Σ is unknown, we replace it with the consistent estimator $\widehat{\Sigma}$.

Corollary 6.1 *If $\widehat{\Sigma}$ is a consistent estimator for Σ, then the CLT still holds, namely*

$$\sqrt{n} \, \widehat{\Sigma}^{-1/2}(\overline{X} - \mu) \xrightarrow{\mathcal{L}} N_d(0, \mathcal{I}_d) \quad \text{as } n \to \infty.$$

Remark 6.1 One may wonder how large n should be in practice to provide reasonable approximations. There is no definite answer to this question. It depends mainly on the problem at hand, i.e. the shape of the distribution and the dimension of X_i. If the X_i are normally distributed, the normality of the sample mean, $\overline{x}$, obtains even from $n = 1$. However, in most situations, the approximation is valid in one-dimensional problems for $n > 50$.

6.3 Copulae

This section describes modelling and measuring the dependency between d rvs using copulae.

Definition 6.1 A d-dimensional copula is a distribution function on $[0, 1]^d$ such that all marginal distributions are uniform on $[0,1]$.

Copulae gained their popularity due to their applications in finance. Sklar (1959) gave a basic theorem on copulae.

Theorem 6.6 (Sklar (1959)) *Let F be a multivariate distribution function with margins $F_1, \ldots, F_d$. Then there exists a copula C such that*

$$F(x_1, \ldots, x_d) = C\{F_1(x_1), \ldots, F_d(x_d)\}, \quad x_1, \ldots, x_d \in \mathbb{R}. \tag{6.11}$$

If the F_i are continuous for $i = 1, \ldots, d$, then C is unique. Otherwise, C is uniquely determined on $F_1(\mathbb{R}) \times \cdots \times F_d(\mathbb{R})$.

Conversely, if C is a copula and $F_1, \ldots, F_d$ are univariate distribution functions, then the function F defined as above is a multivariate distribution function with margins $F_1, \ldots, F_d$.

We can determine the copula of an arbitrary continuous multivariate distribution from the transformation

$$C(u_1, \ldots, u_d) = F\{F_1^{-1}(u_1), \ldots, F_d^{-1}(u_d)\}, \quad u_1, \ldots, u_d \in [0, 1], \qquad (6.12)$$

where the F_i^{-1} are the inverse marginal distribution functions, also referred to as quantile functions. The copula density and the density of the multivariate distribution with respect to the copula are

$$c(u) = \frac{\partial^d C(u)}{\partial u_1, \ldots, \partial u_d}, \quad u \in [0, 1]^d, \qquad (6.13)$$

$$f(x_1, \ldots, x_d) = c\{F_1(x_1), \ldots, F_d(x_d)\} \prod_{i=1}^{d} f_i(x_i), \quad x_1, \ldots, x_d \in \overline{\mathbb{R}}. \quad (6.14)$$

In the multivariate case, the copula function is invariant under monotone transformations.

The estimation and calculation of probability distributions and goodness-of-fit tests are implemented in several R packages: `copula` see Yan (2007), Hofert and Maechler (2011) and Kojadinovic and Yan (2010), `fCopulae` see Wuertz et al. (2009a), `fgac` see Gonzalez-Lopez (2009), `gumbel` see Caillat et al. (2008), HAC see Okhrin and Ristig (2012), `gofCopula` see Trimborn et al. (2015) and `sbgcop` see Hoff (2010). All of these packages have comparative advantages and disadvantages.

The package `sbgcop` estimates the parameters of a Gaussian copula by treating the univariate marginal distributions as nuisance parameters. It also provides a semiparametric imputation procedure for missing multivariate data.

A separate package, `gumbel`, provides functions only for the Gumbel–Hougaard copula. The HAC package focuses on the estimation, simulation and visualisation of Hierarchical Archimedean Copulae (HAC), which are discussed in Sect. 6.3.3. The `fCopulae` package was developed for learning purposes. We recommend using this package for a better understanding of copulae. Almost all the methods in this package, like density, simulation, generator function, etc. can be interactively visualised by changing their parameters with a `slider`. As `fCopulae` is for learning purposes, only the bivariate case is treated, in order to ease the visualisation. The `copula` package tries to cover all possible copula fields. It allows the simulation and fitting of different copula models as well as their testing in high dimensions. As far as we know, this is the only package that deals not only with the copulae, but with the multivariate distributions based on copulae. In contrast to most of the other packages, in `copula` one has to create an object from the classes `copula` or `mvdc` (multivariate distributions constructed from copulae). These classes contain information about the dimension, dependency parameter and margins, in the case of the `mvdc` class. For example, in the following listing, we construct an object that describes a bivariate Gaussian copula with correlation parameter $\rho = 0.75$, with $N(0, 2)$, and $\mathcal{E}(2)$ margins.

```
> n.copula = normalCopula(0.75, dim = 2)
> mvdc.gauss.n.e = mvdc(n.copula,
+    margins      = c("norm","exp"),  # margins of the distribution
+    paramMargins = list(             # parameters of the margins
+       list(mean = 0, sd = 2),       # normal with mu = 0 and sig. = 2
+       list(rate = 2)))              # exponential with rate = 2
```

Using other methods, we can simulate, estimate, calculate and plot the distribution's density function. For copula modelling, we concentrate in this section on the two packages `copula` and `fCopulae`.

6.3.1 Copula Families

We propose three copulae classifications: simple, elliptical and Archimedean; however, there is a zoo of copulae not belonging to these three families.

Simple copulae

Independence and perfect positive or negative dependence properties are often of great interest. If the rvs $X_1, \ldots, X_d$ are stochastically independent, the structure of their relations is given by the product (independence) copula, defined as

$$\Pi(u_1, \ldots, u_d) = \prod_{i=1}^{d} u_i.$$

Two other extremes, representing perfect negative and positive dependencies, are the lower and upper Fréchet–Hoeffding bounds,

$$W(u_1, \ldots, u_d) = \max\left(0, \sum_{i=1}^{d} u_i + 1 - d\right),$$

$$M(u_1, \ldots, u_d) = \min(u_1, \ldots, u_d), \quad u_1, \ldots, u_d \in [0, 1].$$

An arbitrary copula $C(u_1, \ldots, u_d)$ lies between the upper and lower Fréchet–Hoeffding bounds

$$W(u_1, \ldots, u_d) \leq C(u_1, \ldots, u_d) \leq M(u_1, \ldots, u_d).$$

As far as we now, the upper and lower Fréchet–Hoeffdings bounds are not implemented in any package. The reason might be that the lower Fréchet–Hoeffding bound is not a copula function for $d > 2$. Using objects of the class `indepCopula`, one can model the product copula using the functions `dCopula`, `pCopula`, or `rCopula`, described later in this section.

Elliptical copulae

Due to the popularity of the Gaussian and t-distributions in financial applications, the elliptical copulae have an important role. The construction of this type of copula is based on Theorem 6.6 and its implication (6.12). The Gaussian copula and its copula density are given by

$$C_N(u_1, \ldots, u_d, \Sigma) = \Phi_\Sigma\{\Phi^{-1}(u_1), \ldots, \Phi^{-1}(u_d)\},$$

$$c_N(u_1, \ldots, u_d, \Sigma) =$$
$$|\Sigma|^{-1/2} \exp\left[-\frac{\{\Phi^{-1}(u_1), \ldots, \Phi^{-1}(u_d)\}^\top (\Sigma^{-1} - \mathcal{I}_d)\{\Phi^{-1}(u_1), \ldots, \Phi^{-1}(u_k)\}}{2}\right],$$

$$\text{for all } u_1, \ldots, u_d \in [0, 1],$$

where Φ is the distribution function of $N(0, 1)$. Φ^{-1} is the functional inverse of Φ, and Φ_Σ is a d-dimensional normal distribution with zero mean and correlation matrix Σ. The variances of the variables are determined by the marginal distributions.

In the bivariate case, the t-copula and its density are given by

$$C_t(u_1, u_2, \nu, \delta) = \int_{-\infty}^{t_\nu^{-1}(u_1)} \int_{-\infty}^{t_\nu^{-1}(u_2)} \frac{\Gamma\left(\frac{\nu+2}{2}\right)}{\Gamma\left(\frac{\nu}{2}\right)\pi\nu\sqrt{(1-\delta^2)}}$$
$$\times \left\{1 + \frac{x_1^2 - 2\delta x_1 x_2 + x_2^2}{(1-\delta^2)\nu}\right\}^{-\frac{\nu}{2}-1} dx_1 dx_2,$$

$$c_t(u_1, u_2, \nu, \delta) = \frac{f_{\nu\delta}\{t_\nu^{-1}(u_1), t_\nu^{-1}(u_2)\}}{f_\nu\{t^{-1}(u_1)\} f_\nu\{t^{-1}(u_2)\}}, \quad u_1, u_2, \delta \in [0, 1],$$

where δ denotes the correlation coefficient, ν is the number of degrees of freedom, $f_{\nu\delta}$ and f_ν are the joint and marginal t-distributions, respectively, and t_ν^{-1} denotes the quantile function of the t_ν distribution. An in-depth analysis of the t-copula is available in Demarta and McNeil (2004).

The Gaussian or t-copulae can be easily modelled using multivariate normal or t-distribution modelling. Nevertheless, in these packages (see Table 6.2), these copulae are already implemented. As the copula package has the most advanced copula

Table 6.2 Multinormal distribution in R

	fCopulae ($d \geq 2$)	copula ($d \geq 2$)
cdf	dellipticalCopula(type='name')	pCopula(copula)
pdf	pellipticalCopula(type='name')	dCopula(copula)
simul.	rellipticalCopula(type='name')	rCopula(copula)

Note In the fCopulae argument, type can take the values type = {norm, cauchy, t}. For the copula package, the argument of copula should be an object of the classes ellipCopula, normalCopula or tCopula

modelling, we pay special attention to it. As mentioned above, one should first create a `copula` object using `normalCopula`, `tCopula`, or `ellipCopula`.

```
> norm.cop  = normalCopula(       # Gaussian copula
+    param    = c(0.5, 0.6, 0.7), # cor matrix
+    dim      = 3,                 # 3 dimensional
+    dispstr  = "un")              # unstructured cor matrix
> t.cop      = tCopula(           # t copula
+    param = c(0.5, 0.3),         # with params c(0.5, 0.3)
+    dim      = 3,                 # 3 dimensional
+    df       = 2,                 # number of degrees of freedom
+    dispstr  = "toep")           # Toeplitz structure of cor matr.
> norm.cop1 = ellipCopula(        # elliptical family
+    family = "normal",           # Gaussian copula
+    param = c(0.5, 0.6, 0.7),
+    dim = 3, dispstr = "un")      # same as norm.cop
```

The parameter `dispstr` specifies the type of the symmetric positive definite matrix characterising the elliptical copula. It can take the values `ex` for exchangeable, `ar1` for AR(1), `toep` for Toeplitz, and `un` for unstructured. With these objects, one can use the general functions `rCopula`, `dCopula` or `pCopula` for the simulation or calculation of the density or distribution functions.

```
> norm.cop = normalCopula(param = c(0.5, 0.6, 0.7), dim = 3,
+    dispstr = "un")
> # set.seed(2^11-1)             # set the seed, see Chapter 9
> rCopula(n = 3,                 # simulate 3 obs. from a Gaussian cop.
+    copula = norm.cop)
          [,1]        [,2]      [,3]
[1,]  0.6320016  0.3708774 0.7920201
[2,]  0.4009492  0.3540828 0.3455327
[3,]  0.8624083  0.8103213 0.9115499
> dCopula(u = c(0.2,0.5,0.1),    # evaluate the copula density
+    copula = norm.cop)
[1]  1.103629
> pCopula(u = c(0.2,0.5,0.1),    # evaluate the 3D t-copula
+    copula = t.cop)
[1]  0.04190934
```

Plotting the results of these functions is possible for $d = 2$ using the standard `plot`, `persp` and `contour` methods. The following code demonstrates how to use these methods, and the results are displayed in Fig. 6.5.

```
> norm.2d.cop = normalCopula(param = 0.7, dim = 2)
>                                # construct a 2D Gaussian copula
> plot(rCopula(1000, norm.2d.cop)) # scatterplot
> persp(norm.2d.cop, pCopula)    # 3D copula plot
> contour(norm.2d.cop, pCopula)  # copula contour curves
> persp(norm.2d.cop, dCopula)    # 3D plot of the copula density
> contour(norm.2d.cop, dCopula)  # contour curves of the density
```

Using the `mvdc` object on the base of the `copula` objects, one can create a multivariate distribution based on a copula by specifying the parameters of the marginal distributions. The `mvdc` objects can be plotted with `contour`, `persp` or `plot` as well.

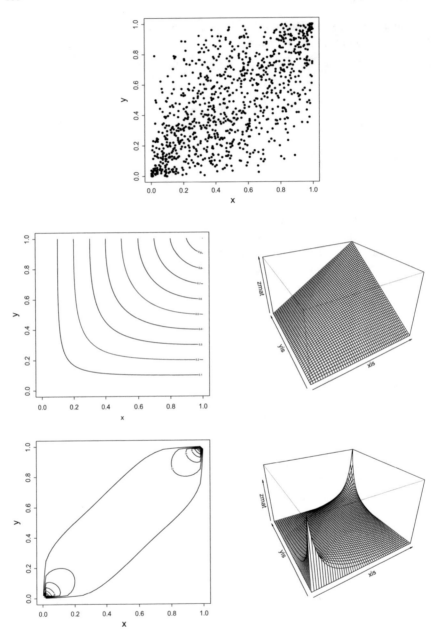

Fig. 6.5 Gaussian copula. *Note* From *top* to *bottom*: scatterplot, distribution and density function for the Gaussian copula with $\rho = 0.7$. **Q** BCS_NormalCopula

```
> n.copula = normalCopula(0.75, dim = 2)
> mvdc.gauss.n.e = mvdc(n.copula, margins = c("norm","exp"),
+   paramMargins = list(list(mean = 0, sd = 2, list(rate = 2)))
> plot(rMvdc(1000, mvdc.gauss.n.e))
> contour(mvdc.gauss.n.e, dMvdc)
> persp(mvdc.gauss.n.e, pMvdc)
```

Using (6.12), one can derive the copula function for an arbitrary elliptical distribution. The problem is, however, that such copulae depend on the inverse distribution functions and these are rarely available in an explicit form, see Sect. 6.2. Therefore, the next class of copulae and their generalisations provide an important flexible and rich family of alternatives to the elliptical copulae.

6.3.2 Archimedean Copulae

As opposed to elliptical copulae, Archimedean copulae are not constructed using Theorem 6.6. Instead, they are related to the Laplace transforms of univariate distribution functions as follows. Let $\mathbb{L}$ denote the class of Laplace transforms which consists of strictly decreasing differentiable functions Joe (1997), i.e.

$$\mathbb{L} = \{\phi : [0; \infty) \to [0, 1] \mid \phi(0) = 1, \ \phi(\infty) = 0; \ (-1)^j \phi^{(j)} \geq 0; \ j = 1, \ldots, \infty\}.$$

The function $C : [0, 1]^d \to [0, 1]$ defined as

$$C(u_1, \ldots, u_d) = \phi\{\phi^{-1}(u_1) + \cdots + \phi^{-1}(u_d)\}, \quad u_1, \ldots, u_d \in [0, 1]$$

is a d-dimensional Archimedean copula, where $\phi \in \mathbb{L}$ is called the *generator of the copula*. It is straightforward to show that $C(u_1, \ldots, u_d)$ satisfies the conditions of Definition 6.1. Some d-dimensional Archimedean copulae are presented below. The use of these copulae in R is similar to that of the elliptical copulae, so we mention only their specific features. The fCopulae package implements the whole list of Archimedean copulae from Nelsen (2006), thus all methods need a parameter type that takes values between 1 and 22.

Frank (1979) copula, $-\infty < \theta < \infty, \ \theta \neq 0.$

The first popular Archimedean copula is the so-called Frank copula, which is the only elliptically contoured Archimedean copula (which is different from elliptical family) for $d = 2$ satisfying

$$C(u_1, u_2) = u_1 + u_2 - 1 + C(1 - u_1, 1 - u_2), \quad u_1, u_2 \in [0, 1].$$

Its generator and copula functions are

$$\phi(x, \theta) = \theta^{-1} \log\{1 - (1 - e^{-\theta})e^{-x}\}, \quad -\infty < \theta < \infty \ \theta \neq 0, \ x \in [0, \infty),$$

$$C_\theta(u_1, \ldots, u_d) = -\frac{1}{\theta} \log \left[1 + \frac{\prod_{j=1}^{d} \{\exp(-\theta u_j) - 1\}}{\{\exp(-\theta) - 1\}^{d-1}} \right].$$

The dependence is maximised when θ tends to infinity and independence is achieved when $\theta = 0$. This family is implemented in the packages `copula` and `fCopulae` under `type` = 5.

Gumbel (1960) copula, $1 \leq \theta < \infty$.

The Gumbel copula is frequently used in financial applications. Its generator and copula functions are

$$\phi(x, \theta) = \exp\{-x^{1/\theta}\}, \quad 1 \leq \theta < \infty, \ x \in [0, \infty),$$

$$C_\theta(u_1, \ldots, u_d) = \exp\left[-\left\{ \sum_{j=1}^{d} (-\log u_j)^\theta \right\}^{\theta^{-1}} \right].$$

Consider a bivariate distribution based on the Gumbel copula with univariate extreme value marginal distributions. Genest and Rivest (1989) showed that this distribution is the only bivariate extreme value distribution based on an Archimedean copula. Moreover, all distributions based on Archimedean copulae belong to its domain of attraction under common regularity conditions. Unlike the elliptical copulae, the Gumbel copula leads to asymmetric contour diagrams. It shows stronger linkages between positive values. However, it also shows more variability and more mass in the negative tail. For $\theta > 1$, this copula allows the generation of a dependence in the upper tail. For $\theta = 1$, the Gumbel copula reduces to the product copula and for $\theta \to \infty$, we obtain the Fréchet–Hoeffding upper bound.

   As mentioned above, apart from the packages `copula` and `fCopulae` (`type` = 4), the package `gumbel`, specially designed for this copula family, allows only exponential or gamma marginal distributions.

Clayton (1978) copula, $-1/(d - 1) \leq \theta < \infty, \ \theta \neq 0$.

The Clayton copula, in contrast to the Gumbel copula, has more mass in the lower tail and less in the upper. The generator and copula function are

$$\phi(x, \theta) = (\theta x + 1)^{-\frac{1}{\theta}}, \quad -1/(d - 1) \le \theta < \infty, \ \theta \ne 0, \ x \in [0, \infty),$$

$$C_\theta(u_1, \ldots, u_d) = \left\{ \left(\sum_{j=1}^{d} u_j^{-\theta} \right) - d + 1 \right\}^{-\theta^{-1}}.$$

The Clayton copula is one of few copulae whose density has a simple explicit form for any dimension

$$c_\theta(u_1, \ldots, u_d) = \prod_{j=1}^{d} \{1 + (j - 1)\theta\} u_j^{-(\theta+1)} \left(\sum_{j=1}^{d} u_j^{-\theta} - d + 1 \right)^{-(\theta^{-1}+d)}.$$

As the parameter θ tends to infinity, the dependence becomes maximal, and as θ tends to zero, we have independence. As θ goes to -1 in the bivariate case, the distribution tends to the lower Fréchet bound. The level plots of the two-dimensional respective densities are given in Fig. 6.6.

6.3.3 Hierarchical Archimedean Copulae

A recently developed flexible method is provided by HAC. The special, so-called fully nested, case of the copula function is

$$
\begin{aligned}
C(u_1, \ldots, u_d) &= \phi_{d-1}\{\phi_{d-1}^{-1} \circ \phi_{d-2}(\ldots [\phi_2^{-1} \circ \phi_1\{\phi_1^{-1}(u_1) + \phi_1^{-1}(u_2)\} \\
&\quad + \phi_2^{-1}(u_3)] + \cdots + \phi_{d-2}^{-1}(u_{d-1})) + \phi_{d-1}^{-1}(u_d)\} \\
&= \phi_{d-1}[\phi_{d-1}^{-1} \circ C(\{\phi_1, \ldots, \phi_{d-2}\})(u_1, \ldots, u_{d-1}) + \phi_{d-1}^{-1}(u_d)]
\end{aligned}
$$

for $\phi_{d-i}^{-1} \circ \phi_{d-j} \in \mathbb{L}^*$, $i < j$, where

$$
\begin{aligned}
\mathbb{L}^* = \{ \omega : [0; \infty) &\to [0, \infty) \mid \omega(0) = 0, \\
\omega(\infty) &= \infty; \ (-1)^{j-1} \omega^{(j)} \ge 0; \ j = 1, \ldots, \infty \}.
\end{aligned}
$$

The HAC defines the whole dependency structure in a recursive way. At the lowest level, the dependency between the first two variables is modelled by a copula function with the generator ϕ_1, i.e. $z_1 = C(u_1, u_2) = \phi_1\{\phi_1^{-1}(u_1) + \phi_1^{-1}(u_2)\}$. At the second level, another copula function is used to model the dependency between z_1 and u_3, etc. Note that the generators ϕ_i can come from the same family, differing only in their parameters. But, to introduce more flexibility, they can also come from different families of generators. As an alternative to the fully nested model, we can consider copula functions with arbitrarily chosen combinations at each copula level. Okhrin et al. (2013) provide several methods for determining the structure of the HAC from the data.

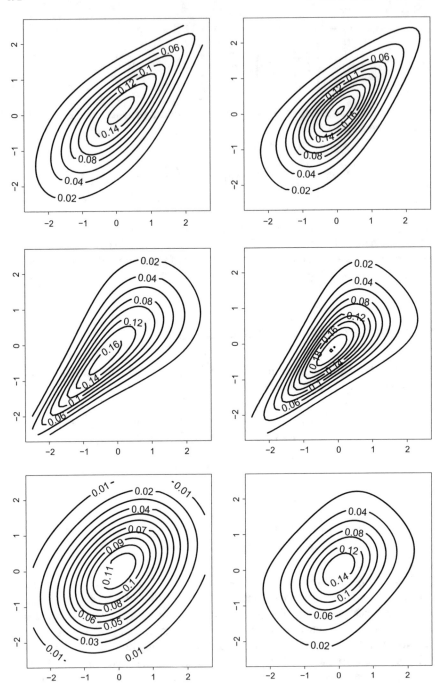

Fig. 6.6 Contour diagrams for (from *top* to *bottom*) the Gumbel, Clayton and Frank copulae with parameter 2 and Normal (*left* column) and t_6 distributed (*right* column) margins.
Q BCS_ArchimedeanContour

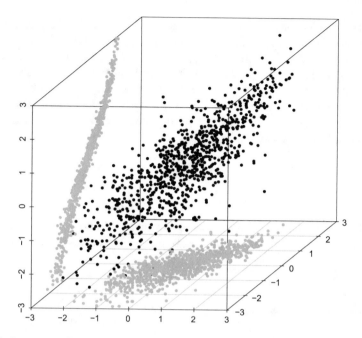

Fig. 6.7 Scatterplot of the HAC-based three-variate distribution. **Q** BCS_HAC

The HAC package provides intuitive techniques for estimating and visualising HAC. In accordance with the naming in the other packages, the functions dHAC, pHAC compute the values of the pdf and cdf, and rHAC generates random vectors. Figure 6.7 presents the scatterplot of the three-dimensional HAC-based distribution.

$$F(x_1, x_2, x_3) = (C_{Gumbel}[C_{Gumbel}\{\Phi(x_1), t_2(x_2); \theta_1 = 2\}, \Phi(x_3); \theta_2 = 10]. \quad (6.15)$$

On the sides of the cube one sees shaded bivariate marginal distribution, that completely differs from each others.

6.3.4 Estimation

Estimating a copula-based multivariate distribution involves both the estimation of the copula parameters θ and the estimation of the margins $F_j(\cdot, \alpha_j)$, $j = 1, \ldots, d$, i.e. all the parameters from the copula and from the margins could be estimated in one step. If we are only interested in the dependency structure, the estimator of θ should be independent of any parametric models for the margins, therefore we distinguish between a parametric and a non-parametric specification of the margins.

In practical applications, however, we are interested in a complete distribution model and, therefore, parametric models for the margins are preferred.

Three commonly used estimation procedures are considered in the following. Let X be a d-dimensional rv with density f and parametric univariate marginal distributions $F_j(x_j; \alpha_j)$, $j = 1, \ldots, d$. Suppose that the copula to be estimated belongs to a given parametric family $\mathcal{C} = \{C_\theta, \theta \in \Theta\}$. For a sample of observations $\{x_i\}_{i=1}^{n}$, $x_i = (x_{1i}, \ldots, x_{di})^\top$ and a vector of parameters $\alpha = (\alpha_1, \ldots, \alpha_d, \theta)^\top \in \mathbb{R}^{d+1}$, the likelihood function is given by

$$L(\alpha; x_1, \ldots, x_n) = \prod_{i=1}^{n} f(x_{1i}, \ldots, x_{di}; \alpha_1, \ldots, \alpha_d, \theta).$$

According to (6.14), the density f can be decomposed into the copula density c and the product of the marginal densities, so that the log-likelihood function can be written as

$$\ell(\alpha; x_1, \ldots, x_n) = \sum_{i=1}^{n} \log c\{F_1(x_{1i}; \alpha_1), \ldots, F_d(x_{di}; \alpha_d); \theta\} + \sum_{i=1}^{n} \sum_{j=1}^{d} \log f_j(x_{ji}; \alpha_j).$$

The vector of parameters $\alpha = (\alpha_1, \ldots, \alpha_d, \theta)^\top$ contains d parameters α_j from the marginals and the copula parameter θ. All these parameters can be estimated in one step through *full maximum likelihood* estimation

$$\tilde{\alpha}_{FML} = (\tilde{\alpha}_1, \ldots, \tilde{\alpha}_d, \tilde{\theta})^\top = \arg\max_{\alpha} \ell(\alpha).$$

Following the standard theory on maximum likelihood estimation, the estimates are efficient and asymptotically normal. However, optimising ℓ with respect to all parameters simultaneously is often computationally demanding.

In the IFM (*inference for margins*) method, the parameters α_j from the marginal distributions are estimated in the first step and used to estimate the dependence parameter θ in the second step. The *pseudo log-likelihood* function

$$\ell(\theta, \hat{\alpha}_1, \ldots, \hat{\alpha}_d) = \sum_{i=1}^{n} \log c\{F_1(x_{1i}; \hat{\alpha}_1), \ldots, F_d(x_{di}; \hat{\alpha}_d); \theta\},$$

is maximised over θ to get the dependence parameter estimate $\hat{\theta}$. A detailed discussion of this method is to be found in Joe and Xu (1996). Note that this procedure does not lead to efficient estimators, but, as argued by Joe (1997), the loss in efficiency should be modest. The advantage of the inference for margins procedure lies in the dramatic reduction of the computational complexity, as the estimation of the margins is disentangled from the estimation of the copula. As a consequence, all R packages use the separate estimation of the copula and its margins and, therefore, they focus only on the optimisation of the copula parameter(s).

In the CML (*canonical maximum likelihood*) method, the univariate marginal distributions are estimated through some non-parametric method $\hat{F}$ as described in Sect. 5.1.2. The asymptotic properties of the multistage estimators of θ do not depend explicitly on the type of the non-parametric estimator, but on its convergence properties. For the estimation of the copula, one should normalise the empirical cdf not by n but by $n+1$:

$$\hat{F}_j(x) = \frac{1}{n+1} \sum_{i=1}^{n} I(x_{ji} \leq x).$$

The copula parameter estimator $\hat{\theta}_{CML}$ is given by

$$\hat{\theta}_{CML} = \arg\max_{\theta} \sum_{i=1}^{n} \log c\{\hat{F}_1(x_{1i}), \dots, \hat{F}_d(x_{di}); \theta\}.$$

Notice that the first step of the IMF and CML methods estimates the marginal distributions. After the estimation of the marginal distributions, a *pseudosample* $\{u_i\}$ of observations transformed to the unit d-cube is obtained and used for the *copula* estimation. As in the IFM, the semiparametric estimator $\hat{\theta}$ is asymptotically normal under suitable regularity conditions.

In the two-dimensional case $d = 2$, one often uses the generalised method of moments, since there is a one to one relation between the bivariate copula parameter and Kendall's τ or Spearman's ρ_S. For example, for Gumbel copulae, $\tau = 1 - \frac{1}{\theta}$, and for Gaussian copulae, $\tau = \frac{2}{\pi} \arcsin \rho$. One estimates this measure as (6.7) or (6.6) and subsequently converts it to θ.

Estimation of the different copula models is implemented in a variety of packages, such as copula, fCopulae, gumbel and HAC. The gumbel package implements all methods of estimation for the Gumbel copula. The package fCopulae deals only with copula functions with uniform margins, and the estimation is provided through maximum likelihood. It estimates the parameters for all the copula families in Nelsen (2006). The package copula is of the highest interest, since almost all estimation methods for the estimation of multivariate copula-based distributions, or just the copula function, are implemented in it. To estimate a parametric copula C, one uses the fitCopula function, which among other parameters needs the parameter method, indicating the method that should be used in the estimation. The parameter method can be either ml (maximum likelihood), mpl (maximum pseudo-likelihood), itau (inversion of Kendall's tau), or irho (inversion of Spearman's rho). The default method is mpl. In the following listing, we present the estimation of the Gumbel copula parameter using different methods.

```
> # set.seed(11)                              # set seed, see Chapter 9
> Gc = gumbelCopula(3, dim = 2)               # Gumbel copula
> n  = 200                                    # sample size
> x  = rCopula(n, copula = Gc)                # true observations
> u  = apply(x, 2, rank) / (n + 1)            # pseudo-observations
> fitCopula(Gc, u, method ="itau")@estimate   # inverting Kendall's tau
[1] 2.57772
> fitCopula(Gc, u, method ="irho")@estimate   # inv. Spearman's rho
[1] 2.600561
> fitCopula(Gc, u, method ="mpl")@estimate    # maximum pseudo-lik.
[1] 2.657331
> fitCopula(Gc, x, method ="ml")@estimate     # maximum likelihood
[1] 2.657331
```

Similarly, using the `fitMvdc` method, one can estimate the whole multivariate copula-based distribution together with its margins.

Chapter 7
Regression Models

Everything must be made as simple as possible. But not simpler.

— Albert Einstein

Regression models are extremely important in describing relationships between variables. Linear regression is a simple, but powerful tool in investigating linear dependencies. It relies, however, on strict distributional assumptions. Nonparametric regression models are widely used, because fewer assumptions about the data at hand are necessary. At the beginning of every empirical analysis, it is better to look at the data without assumptions about the family of distributions. Nonparametric techniques allow describing the observations and finding suitable models.

7.1 Idea of Regression

Regression models aim to find the most likely values of a dependent variable Y for a set of possible values $\{x_i\}$, $i = 1, \ldots, n$ of the explanatory variable X

$$y_i = g(x_i) + \varepsilon_i, \quad \varepsilon_i \sim F_\varepsilon,$$

where $g(x) = \mathsf{E}(Y|X = x)$ is an arbitrary function which is included in the model with the intention of capturing the mean of the process that corresponds to x. The natural aim is to keep the values of the ε_i as small as possible.

Parametric models assume that the dependence of Y on X can be fully explained by a finite set of parameters and that F_ε has a prespecified form with parameters to be estimated.

Nonparametric methods do not assume any form: neither for g nor for F_ε, which makes them more flexible than the parametric methods. The fact that nonparametric techniques can be applied where parametric ones are inappropriate prevents the nonparametric user from employing a wrong method. These methods are particularly

© Springer International Publishing AG 2017
W.K. Härdle et al., *Basic Elements of Computational Statistics*,
Statistics and Computing, DOI 10.1007/978-3-319-55336-8_7

useful in fields like quantitative finance, where the underlying distribution is in fact unknown. However, as fewer assumptions can be exploited, this flexibility comes with the need for more data. A detailed introduction to nonparametric techniques can be found in Härdle et al. (2004).

7.2 Linear Regression

The simplest relationship between quantitative variables, an explained variable Y and explanatory variables $X_1, \ldots, X_p$ (also called *regressors*) is the linear one, namely the function $g(\cdot)$ is linear. It is of interest to study 'how Y varies with changes in $X_1, \ldots, X_p$', or precisely, the *causal* (ceteris paribus) relationship between variables. The model, which is assumed to hold in the population, is written as

$$Y = \beta_0 + \beta_1 X_1 + \cdots + \beta_p X_p + \varepsilon.$$

The variable ε is called the error term and represents all factors other than $X_1, \ldots, X_d$ that affect Y. Let $y = \{y_i\}_{i=1}^n$ be a vector of the response variables and $\mathcal{X} = \{x_{ij}\}_{i=1,\ldots,n;\ j=1,\ldots,p}$ be a data matrix of p explanatory variables. In many cases, a constant is included through $x_{i1} = 1$ for all i in this matrix. The resulting data matrix is denoted by $\mathcal{X} = \{x_{ij}\}_{i=1,\ldots,n;\ j=1,\ldots,p+1}$. The aim is to find a good linear approximation of y using linear combinations of covariates

$$y = \mathcal{X}\beta + \varepsilon,$$

where ε is the vector of errors. To estimate β, the following least squares optimisation has to be solved:

$$\hat{\beta} = \underset{\beta \in \Theta}{\operatorname{argmin}} \|y - \mathcal{X}\beta\|^2 = \underset{\beta \in \Theta}{\operatorname{argmin}} (y - \mathcal{X}\beta)^\top (y - \mathcal{X}\beta) = \underset{\beta \in \Theta}{\operatorname{argmin}} \varepsilon^\top \varepsilon, \quad (7.1)$$

where Θ denotes the parameter space.

Applying first-order conditions, it can be shown that the solution of (7.1) is

$$\hat{\beta} = (\mathcal{X}^\top \mathcal{X})^{-1} \mathcal{X}^\top y. \quad (7.2)$$

This estimator is called the (ordinary) least squares (OLS) estimator. Under the following conditions, the OLS estimator $\hat{\beta}$ is by the Gauss–Markov theorem the best linear unbiased estimator (BLUE).

1. The model is linear in the parameter vector β;
2. $\mathsf{E}(\varepsilon) = 0$;
3. the covariance matrix of the errors ε is diagonal and given by $\Sigma = \sigma^2 \mathcal{I}_n$;
4. $\mathcal{X}$ is a deterministic matrix with full column rank.

If these conditions are fulfilled, then OLS has the smallest variance in the class of all linear unbiased estimators, with $\mathsf{E}(\hat{\beta}) = \beta$ and $\mathsf{Var}(\hat{\beta}) = \sigma^2(\mathcal{X}^\top \mathcal{X})^{-1}$.

Additional assumptions are required to develop further inference about $\hat{\beta}$. Under a normality assumption, $\varepsilon \sim N(0, \sigma^2 \mathcal{I}_n)$, the estimator $\hat{\beta}$ has a normal distribution, i.e. $\hat{\beta} \sim N\{\beta, \sigma^2(\mathcal{X}^\top \mathcal{X})^{-1}\}$. In practice, the error variance σ^2 is often unknown, but can be estimated by

$$\widehat{\sigma}^2 = \frac{1}{n-(p+1)}(y-\hat{y})^\top(y-\hat{y}),$$

where $\hat{y}$ are the predictors of y given by $\hat{y} = \mathcal{X}\hat{\beta}$ and $(p+1)$ is the number of explanatory variables including the intercept.

Given the structure of the matrix $\mathsf{Var}(\hat{\beta})$, the standard deviation of a single estimator $\hat{\beta}_j$ can be written as

$$\widehat{\sigma}^2(\hat{\beta}_j) = \frac{\sum_{i=1}^n \hat{\varepsilon}_i^2}{n-(p+1)} = \frac{\hat{\varepsilon}^\top \hat{\varepsilon}}{n-(p+1)} = \widehat{\sigma}^2(\mathcal{X}^\top \mathcal{X})_{jj}^{-1},$$

where $(\mathcal{X}^\top \mathcal{X})_{jj}$ is the j-th diagonal element of the matrix. It can be shown that $\hat{\beta}_j$ and $\hat{\sigma}^2(\hat{\beta}_j)$ are statistically independent.

The distributional property of $\hat{\beta}$ is used to form tests and build confidence intervals for the vector of parameters β. Testing the hypothesis $H_0 : \beta_j = 0$ is analogous to the one-dimensional t-test with the test statistics given by

$$t = \hat{\beta}_j/\widehat{\sigma}(\hat{\beta}_j). \tag{7.3}$$

Under H_0, the test statistic (7.3) follows the $t_{n-(p+1)}$ distribution. For further reading we refer to Greene (2003), Härdle and Simar (2015) and Wasserman (2004).

Additionally one can test whether all independent variables have no effect on the dependent variable

$$H_0 : \beta_1 = \ldots = \beta_p = 0 \quad \text{vs} \quad H_1 : \beta_k \neq 0, \quad for\ at\ least\ one\ k = 1, \ldots, p$$

In order to decide on the rejection of the null hypothesis, the residual sum of squares *RSS* serves as a measure. According to the restrictions, under the null hypothesis the test compares the *RSS* of the reduced model $SS(reduced)$, in which the variables listed in H_0 are dropped, with the *RSS* of an unrestricted model $SS(full)$, in which all variables are included. In general, $SS(reduced)$ is greater than or equal to $SS(full)$ because the OLS estimation of the restricted model uses fewer parameters. The question is whether the increase of the *RSS* in moving from the unrestricted model to the restricted model is large enough to ensure the rejection of the null hypothesis. Therefore, the F-statistic is used, which addresses the difference between $SS(reduced)$ and $SS(full)$:

$$F = \frac{SS(reduced) - SS(full)/df(r) - df(f)}{SS(full)/df(f)}. \quad (7.4)$$

Under the null hypothesis, the statistic (7.4) follows the distribution $F_{df(r)-df(f),df(r)}$ (see Sect. 4.4.3), where $df(f)$ and $df(r)$ denote the number of degrees of freedom under the unrestricted model and the restricted model $(df(f) = n - p - 1$ and $df(r) = n - 1)$. Based on this F-distribution, the critical region can be chosen in order to reject or not reject the null hypothesis.

7.2.1 Model Selection Criteria

Even in the case of well-fitted models, it is not an easy task to select the best model from a set of alternatives. Usually, one looks at the coefficient of determination R^2 or *adjusted* R^2. These values measure the 'goodness of fit' of the regression equation. They represent the percentage of the total variation of the data explained by the fitted linear model. Consequently, higher values indicate a 'better fit' of the model, while low values may indicate a poor model. R^2 is given by

$$R^2 = 1 - \frac{\|y - \hat{y}\|^2}{\|y - \overline{y}\|^2}, \quad (7.5)$$

with $R^2 \in [0, 1]$. It is important to know, that R^2 always increases with the number of explanatory variables added to the model even if they are irrelevant.

The adjusted R^2 is a modification of (7.5), which considers the number of explanatory variables used, and is given by

$$R^2_{adj} = 1 - (1 - R^2)\frac{n-1}{n-(p+1)-1}. \quad (7.6)$$

Note that R^2_{adj} can be negative. However, the coefficients of determination (7.5) and (7.6) are not always the best criteria for choosing the model. Other popular criteria to choose the regression model are Mallows' C_p, the Akaike Information Criterion (AIC) and the Bayesian Information Criterion (BIC).

Mallows' C_p is a model selection criterion which uses the residual sum of squares, but penalises for the number of unknown parameters like R^2_{adj}. It is given by

$$C_p = \frac{\|y - \hat{y}\|^2}{\|y - \overline{y}\|^2} - n + 2(p+1).$$

AIC uses the maximum log-likelihood and is defined as

$$\text{AIC} = n \log \hat{\sigma}^2 + 2(p+1),$$

where p is the number of parameters and $\widehat{\sigma}^2$ an estimate for the variance maximising L a likelihood function. The second term is a penalty, as in Mallows' C_p.

The last information criterion discussed here is the BIC, defined as

$$\text{BIC} = n \log \widehat{\sigma}^2 + \log(n)(p + 1).$$

There is no rule of thumb determining which criterion to use. In small samples all criteria give similar results. Since BIC has a larger penalty for $n \geq 3$ than AIC, it will have a tendency to select more parsimonious models.

7.2.2 Stepwise Regression

Stepwise regression builds the model from a set of candidate predictor variables by entering and removing predictors in a stepwise manner. One can perform forward or backward stepwise selection using the `step` function or `stepAIC` from the `MASS` package. Both functions perform stepwise model selection by exact AIC. The `stepAIC` function is preferable, because it is applicable to more model types, e.g. nonlinear regression models, apart from the linear model while providing the same options. Forward selection starts by choosing the independent variable which explains the most variation in the dependent variable. It then chooses the variable which explains most of the remaining residual variation and recalculates the regression coefficients. The algorithm continues until no further variable significantly explains the remaining residual variation. Another similar selection algorithm is backward selection, which starts with all variables and excludes the most insignificant variables one at a time, until only significant variables remain. A combination of the two algorithms performs forward selection, while dropping variables which are no longer significant after the introduction of a new variable.

The `stepAIC()` function requires a number of arguments. The argument `k` is a multiple of the number of degrees of freedom used for the penalty. If `k = 2` the original AIC is applied, `k = log(n)` is equivalent to BIC. The direction of the stepwise regression can be chosen as well, setting it to forward, backward or both. If `trace = 1`, it will return every model it goes over as well as the coefficients of the final model. `scope` gives the range of the included predictors, while `lower` and `upper` specify the minimal and maximal number of models the stepwise procedure may go over.

The nutritional database on US cereals introduced in Venables and Ripley (1999) provides a good illustration of MLR. The `UScereal` data frame from package `MASS` is from the 1993 ASA Statistical Graphics Exposition. The data have been normalised to a portion size of one American cup and among other contain information on: `mfr` (Manufacturer, represented by its first initial), `calories` (number of calories per portion), `protein` (grams of protein per portion), `fat` (grams of fat per portion), `carbo` (grams of complex carbohydrates per portion) and sugars (grams of sugars per portion). The analysis is restricted to the dependence between calories and protein, fat, carbohydrates and sugars, which can be expressed by the function

$$CALORIES = \beta_1 \cdot INTERCEPT + \beta_2 \cdot PROTEIN + \beta_3 \cdot FAT$$
$$+ \beta_4 \cdot CARBO + \beta_5 \cdot SUGARS + \varepsilon.$$

First, it is necessary to load the package which contains the dataset.

```
> data("UScereal", package ="MASS")
```

Function `lm()` estimates the coefficients of a multiple linear regression, calculates the standard errors of the estimators and tests the regression coefficients' significance.

```
> fit = lm(calories ~ protein + fat + carbo + sugars,
+    data = UScereal) # fit the regression model
```

The resulting fitted model is an object of a class `lm`, for which the function `summary()` shows the conventional regression table. The part `Call` states the applied model.

```
> summary(fit)            # show results
Call:
lm(formula = calories ~ protein + fat + carbo + sugars,
data = UScereal)
```

The next part of the output provides the minimum, maximum and empirical quartiles of the residuals.

```
Residuals:                    # output ctnd.
    Min     1Q  Median     3Q     Max
-20.177  -4.693   1.419  4.940  24.758
```

The last part shows the estimated $\hat{\beta}$ with the corresponding standard errors, t-statistics and associated p-values. The measures of goodness of fit, R^2 and adjusted R^2 as discussed in Sect. 7.2.1, and the results of an F-test are given as well. The F-test tests the null hypothesis that all regression coefficients (excluding the constant) are simultaneously equal to 0.

```
Coefficients:                      # output ctnd.
              Estimate Std. Error t value Pr(>|t|)
(Intercept) -18.7698      3.5127  -5.343 1.49e-06 ***
protein       4.0506      0.5438   7.449 4.28e-10 ***
fat           8.8589      0.7973  11.111 3.41e-16 ***
carbo         4.9247      0.1587  31.040  < 2e-16 ***
sugars        4.2107      0.2116  19.898  < 2e-16 ***
---
Signif. codes:  0 "***" 0.001 "**" 0.01 "*" 0.05 "."

Residual standard error: 8.862 on 60 degrees of freedom
Multiple R-squared: 0.9811,    Adjusted R-squared: 0.9798
F-statistic: 778.6 on 4 and 60 DF,  p-value: < 2.2e-16
```

In the given example, all four variables are statistically significant, i.e. all p-values are smaller than 0.05, which is commonly used as threshold. The coefficients can be assessed via the command `fit$coef`. The interpretation of the coefficients is quite intuitive. One may know from a high school chemistry course that 1 g of carbohydrates or proteins contains 4 calories and 1 g of fat gives nine calories. In order to investigate the model closely, four diagnostic plots are constructed. The layout command is used to put all graphs in one figure.

```
> layout(matrix(1:4, 2, 2))        # plot 4 graphics in 1 figure
> plot(fit)                        # depict 4 diagnostic plots
```

The upper left plot in Fig. 7.1 shows the residual errors plotted against their fitted values. The residuals should be randomly distributed around the horizontal axis. There should be no distinct trend in the distribution of points. A nonparametric regression of the residuals is added to the plot, which should, in an ideal model, be close to the horizontal line $y = 0$. Unfortunately, this is not the case in the example, possibly due to outliers (Grappe-Nuts, Quaker Oat Squares and Bran Chex). Potential outliers are always named in the diagnostic plots.

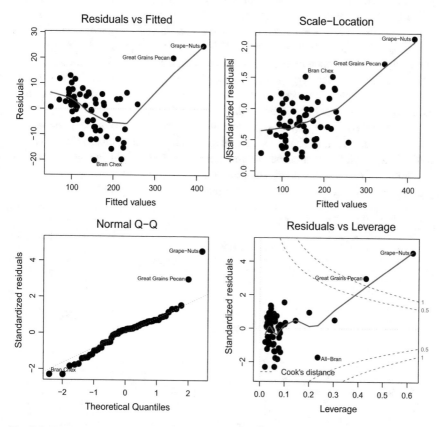

Fig. 7.1 Diagnostic plots for multiple linear regression. Q BCS_MLRdiagnostic

The upper right plot in Fig. 7.1 presents the scale-location, also called spread-location. It shows the square root of the standardised residuals as a function of the fitted values and is a useful tool to verify if the standard deviation of the residuals is constant as a function of the dependent variable. In this example, the standard deviation increases with the number of calories.

The lower left plot in Fig. 7.1 is a Q-Q plot. In a Q-Q plot, quantiles of a theoretical distribution (Gaussian in this case) are plotted against empirical quantiles. If the data points lie close to the diagonal line there is no reason to doubt the assumed distribution of the errors (e.g. Gaussian).

Finally, the lower right plot shows each point's leverage, which is a measure of its importance in determining the regression result. The leverage of the i-th observation is the i-th diagonal element of the matrix $\mathcal{X}(\mathcal{X}^\top \mathcal{X})^{-1}\mathcal{X}^\top$. It always takes values between 0 and 1 and shows the influence of the given observation on the overall modelling results and particularly on the size of the residual. Superimposed on the plot are contour lines for Cook's distance, which is another measure of the importance of each observation to the regression, showing the change in the predicted values when that observation is excluded from the dataset. Smaller distances mean that this observation has little effect on the regression. Distances larger than 1 are suspicious and suggest the presence of possible outliers or a poor model. For more details, we refer to Cook and Weisberg (1982). In the given regression model, some possible outliers are observed. It makes sense to have a closer look at these observations and either exclude them or experiment with other model specifications.

As mentioned above, adjusted R^2 is a widely used measure of the goodness of fit. In the given example, the model seems to explain the variability in the data quite well, the R^2_{adj} is 0.9798. However, a similar goodness of fit might be obtained using a smaller set of regressors. An investigation of this question using stepwise regression procedure shows that no regressor can be removed from the model.

```
> require(MASS)
> stepAIC(fit, direction ="both") # stepwise regression using AIC
Start:  AIC = 288.43
calories ~ protein + fat + carbo + sugars

          Df Sum of Sq   RSS    AIC
<none>                  4712 288.43
- protein  1      4358  9070 328.99
- fat      1      9694 14406 359.07
- sugars   1     31092 35804 418.24
- carbo    1     75664 80376 470.81


Call:
lm(formula = calories ~ protein + fat + carbo + sugars,
data = UScereal)

Coefficients:
(Intercept)        protein          fat        carbo       sugars
    -18.770          4.051        8.859        4.925        4.211
```

Fig. 7.2 BIC for all subsets regression. Q BCS_MLRleaps

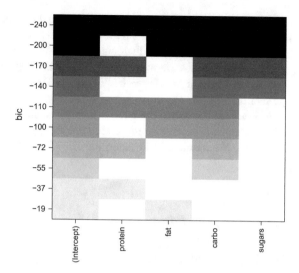

A possible drawback of stepwise regression is that once the variable is included (excluded) in the model, it remains there (or is eliminated) for all remaining steps. Thus, it is a good idea to perform stepwise regression in both directions in order to look at all the possible combinations of explanatory variables. It is possible to perform an all subsets regression using the function `regsubsets` from the package `leaps`. By plotting `regsubset` object one obtain a matrix, on which with dark colour are highlighted models with larger BIC (Fig. 7.2).

```
> require("leaps")
> sset = regsubsets(calories ~ protein + fat + carbo + sugars,
+     data = UScereal, nbest = 3)   # fit lm to all subsets
> plot(sset)
```

7.3 Nonparametric Regression

The general idea of regression analysis is to find a reasonable relation between two variables X and Y. For n realisations $\{(x_i, y_i)\}_{i=1}^{n}$, the relation can be modelled by

$$y_i = g(x_i) + \varepsilon_i, \quad i = 1, ..., n, \tag{7.7}$$

where X is our explanatory variable, Y is the explained variable, and ε is the noise.

A parametric estimation would suggest $g(x_i) = g(x_i, \theta)$, therefore estimating g would result in estimating θ and using $\hat{g}(x_i) = g(x_i, \hat{\theta})$. In contrast, nonparametric regression allows g to have any shape, in the sense that g need not belong to a set of defined functions. Nonparametric regression provides a powerful tool by allowing wide flexibility for g. It avoids a biased regression, and might be a good starting point for a parametric regression if no 'a priori' shape of g is known. It is also a reliable way to spot outliers within the regression.

There are two cases to distinguish, depending on the assumptions. Assuming that the observations are independent and identically distributed, we can write the regression function as $g(x) = E(Y|X = x)$. This is the so-called random design problem.

The second problem, the fixed design problem, is in some ways better for the statistician, as it provides a more accurate estimator of the regression function. In this setup, X is assumed to be deterministic. We keep the hypothesis that the $\{\varepsilon_i\}$ are i.i.d.

7.3.1   General Form

For (7.7), the following must hold:

1. $E(\varepsilon) = 0$ and $\text{Var}\,\varepsilon = \sigma^2$.
2. $g(x) \approx y_i$, for $x \approx x_i$.
3. If x is 'far' from x_j, we want (x_j, y_j) to not interact with $g(x)$.

Therefore one has to build a function as a finite sum of g_i, where g_i is related to the point (x_i, y_i). Without loss of generality,

$$\hat{g}(x) = \sum_{i=1}^{n} g_i(x), \quad g_i(x) \approx \begin{cases} y_i, & \text{if } x \text{ is close to } x_i \\ 0, & \text{else.} \end{cases}$$

This can be written as a weighted sum or local average

$$\hat{g}(x) = \sum_{i=1}^{n} w_i(x) y_i, \tag{7.8}$$

with the weight $w_i(x)$ for each x_i.

This form is quite appealing, because it is similar to the solution of the least squares minimization problem

$$\min_{\theta \in \mathbb{R}} \sum_{i=1}^{n} w_i(x)(\theta - y_i)^2,$$

which is solved for θ. This means that finding a local average is the same as finding a locally weighted least squares estimate. For more details on the distinction between local polynomial fitting and kernel smoothing, see Müller (1987).

A method similar to the histogram divides the set of observations into bins of size h and computes the mean within each bin by

Fig. 7.3 Nonparametric regression of daily DAX log-returns by daily FTSE log-returns, with a uniform kernel and $h = 0.001$.
Q BCS_KernelSmoother

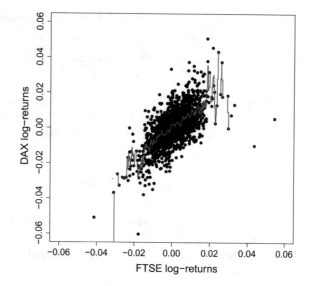

$$\hat{g}(x) = \frac{\sum_{i=1}^{n} I\{|x - x_i| < h/2\} y_i}{\sum_{i=1}^{n} I\{|x - x_i| < h/2\}}. \tag{7.9}$$

This is implemented in R via the function `ksmooth()` choosing the parameter `kernel ='box'`. As an example (Fig. 7.3), we choose to regress the daily log-returns of the DAX index on the daily log-returns of the FTSE index with $h = 0.001$ (how the returns are computed see Sect. 5.3.1).

```
> r.dax      = diff(log(EuStockMarkets[, 1]))   # daily DAX log returns
> r.ftse     = diff(log(EuStockMarkets[, 4]))   # daily FTSE log returns
> r.daxhat = ksmooth(x = r.ftse, y = r.dax,     # kernel regression
+     kernel   ="box",                           # type of kernel
+     bandwidth = 0.001)                          # bandwidth
> plot(r.ftse, r.dax)                            # plot observations
> lines(r.daxhat, col ="red3")                   # plot kernel regression
```

The function `ksmooth()` returns the fitted values of the DAX log-returns and the respective FTSE log-returns. This estimator is a special case of a wider family of estimators, which is the topic of the following section.

7.3.2 Kernel Regression

As with the density estimation, we can bring into play the kernel functions for the weights we need in (7.8). Recall the density estimator

$$\hat{f}(x) = \frac{1}{nh} \sum_{i=1}^{n} K\left(\frac{x - x_i}{h}\right).$$

We then have the general estimator $\hat{g}(x)$ related to the bandwidth h and to a kernel K

$$\hat{g}(x) = \frac{\sum_{i=1}^{n} K(\frac{x-x_i}{h})y_i}{\sum_{i=1}^{n} K(\frac{x-x_i}{h})}. \tag{7.10}$$

This generalisation of the kernel regression estimator is attributed to Nadaraya (1964) and Watson (1964), and is the so-called *Nadaray–Watson* estimator. If one knows the actual density f of X, then (7.10) reduces to

$$\hat{g}(x) = \sum_{i=1}^{n} \frac{K(\frac{x-x_i}{h})y_i}{f(x)}.$$

These estimators are of the form of (7.8), with weights equal to $w_i(x) = K\left(\frac{x-x_i}{h}\right)/$ $f(x)$ and $w_i(x) = K\left(\frac{x-x_i}{h}\right)/\sum_{i=1}^{n} K\left(\frac{x-x_i}{h}\right)$.

In the following, some regressions with different kernels and different bandwidths are computed.

```
> r.dax    = diff(log(EuStockMarkets[, 1]))   # daily DAX log returns
> r.ftse   = diff(log(EuStockMarkets[, 4]))   # daily FTSE log returns
> n        = length(r.dax)                     # sample size
> h        = c(0.1, n^-1, n^-0.5)              # bandwidths
> Color    = c("red3","green3","blue3")        # vector for colors
> # kernel regression with uniform kernel
> r.dax.un = list(h1 = NA, h2 = NA, h3 = NA)   # list for results
> for(i in 1:3){
+    r.dax.un[[i]] = ksmooth(x = r.ftse,        # independent variable
+       y          = r.dax,                     # dependent variable
+       kernel     = "box",                     # use uniform kernel
+       bandwidth = h[i])                       # h = 0.1, n^-1, n^-0.5
+ }
> plot(x = r.ftse, y = r.dax)                   # scatterplot for data
> for(i in 1:3){
+    lines(r.dax.un[[i]], col = Color[i])       # regression curves
+ }
> # kernel regression with normal kernel
> r.dax.no = list(h1 = NA, h2 = NA, h3 = NA)   # list for results
> for(i in 1:3){
+    r.dax.no[[i]] = ksmooth(x = r.ftse,        # independent variable
+       y          = r.dax,                     # dependent variable
+       kernel     = "normal",                  # use normal kernel
+       bandwidth = h[i])                       # h = 0.1, n^-1, n^-0.5
+ }
> plot(x = r.ftse, y = r.dax)                   # scatterplot for data
> for(i in 1:3){
+    lines(r.dax.no[[i]], col = Color[i])       # regression curves
+ }
```

As previously noted in Sect. 5.1.4, the choice of the bandwidth h is crucial for the degree of smoothing, see Fig. 7.4. In Fig. 7.5 the Gaussian kernel $K(x) = \varphi(x)$ is used with the same bandwidths. The kernel determines the shape of the estimator $\hat{g}$, which is illustrated in Figs. 7.4 and 7.5.

Fig. 7.4 Nonparametric
regression of daily DAX
log-returns by daily FTSE
log-returns, using uniform
kernel. The plot shows the
density estimates for
different bandwidths:
$h = 0.1, h = n^{-1}$ and
$h = n^{-1/2}$.
Q BCS_UniformKernel

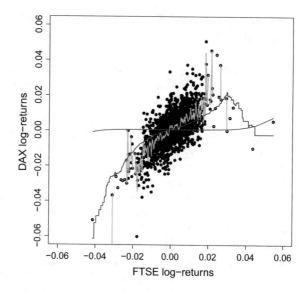

Fig. 7.5 Nonparametric
regression of daily DAX
log-returns by daily FTSE
log-returns, using Gaussian
kernel. The plot shows the
density estimates for
different bandwidths:
$h = 0.1, h = n^{-1}$ and
$h = n^{-1/2}$.
Q BCS_GaussianKernel

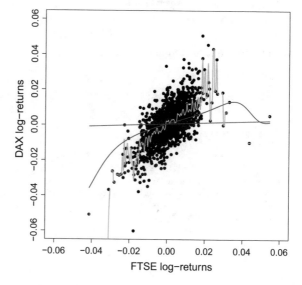

7.3.3 *k-Nearest Neighbours (k-NN)*

The idea of nonparametric regression is to build a local average of $\{y_i\}$ to obtain a
reasonable value for $g(x)$, where x is close to some $\{x_i\}$. Instead of using a kernel
regression, one could build subsets for the domain of X around x and construct the

local average for each different subset. In other words, if $\{S_i\}$ is a family of disjoint subsets of the domain S of X and $S = \cup_{i \in I} S_i$, with $I = \{1, \ldots, p\}$, then one uses

$$\hat{g}(x) = \begin{cases} \frac{1}{|\{j, x_j \in S_i\}|} \sum_{j, x_j \in S_i} y_j, & \text{if } \exists i, x \in S_i, \\ 0, & \text{else.} \end{cases} \tag{7.11}$$

For instance, if one picks $S_i(h) = \{x, |x - x_i| < h/2\}$, then this is simply the uniform kernel regressor with bandwidth h. An alternative is to choose S_i such that the k-nearest observations x_i to x, in terms of the Euclidean distance, are selected. This avoids the regressor's being equal to 0, and has an intuitive foundation. Since the estimator is computed with the k-nearest points, it is less sensitive to outliers in the dataset. However, there can be a lack of accuracy when k is large compared to n (the sample size). The estimator will give the same weight to neighbours that are close and far away. This problem is less severe with a larger number of observations, or in the case of the 'fixed design' problem, where x_i is selected by the user. In the case of a small number of observations, one can also compensate for this lack of consistency by combining this method with a kernel regression. R includes an implementation of the k-NN algorithm for dependent variables Y. The function in R is knn() from package class.

Consider again the DAX log-returns from the EuStockMarkets dataset. The probability of having positive DAX log-returns conditional on the FTSE, CAC and SMI log-returns, is computed in the following.

```
> require(class)
> k     = 20                                   # neibourghs
> data  = diff(log(EuStockMarkets))            # log-returns
> size  = (dim(data)[1] - 9):dim(data)[1]      # last ten obs.
> train = data[-size, -1]                      # training set
> test  = data[size, -1]                       # testing set
> cl    = factor(ifelse(data[-size, 1] < 0,    # returns as factor
+   "decrease","increase"))
> tcl   = factor(ifelse(data[size, 1] < 0,     # true classification
+   "decrease","increase"))
> pcl   = knn(train, test, cl, k, prob = TRUE) # predicted returns
> pcl
 [1] decrease decrease decrease decrease
     increase decrease decrease increase
     decrease increase
attr(, "prob")
 [1] 0.95 0.90 0.95 0.90 1.00 1.00 0.95 1.00 0.90 1.00
Levels: decrease increase
> tcl == pcl                                    # validation
 [1] TRUE TRUE TRUE TRUE TRUE TRUE TRUE TRUE TRUE TRUE
```

The predicted classifications for the DAX log-returns fit perfectly the actual classifications. All predicted probabilities are at least 0.90.

This simple call to the k-NN estimator allows only for one set of y_i. However, manually coding a k-NN function does not require much depth in the reasoning if we keep it simple. One may write the following code:

```
> r.dax    = diff(log(EuStockMarkets[,1]))    # log-returns of DAX
> r.ftse   = diff(log(EuStockMarkets[,4]))    # log returns of FTSE
> knn.reg = function(x, xis, yis, k){         # function for neighbours
+ knn.reg = rep(0, times = length(x))         # empty object
+   for (i in 1:length(x)){                   # loop over length
+     distances = order(abs(x[i] - xis))      # order distances
+     knn.reg[i] = mean(yis[distances][1:k])  # mean over neighbours
+   }
+   knn.reg
+ }
> # functions of regressions for different amount of neighbours
> knn.reg.k1 = function(x)knn.reg(x, r.ftse, r.dax, 10)
> knn.reg.k2 = function(x)knn.reg(x, r.ftse, r.dax, 250)
> knn.reg.k3 = function(x)knn.reg(x, r.ftse, r.dax, 1)
> plot(r.ftse, r.dax)                         # plot scatterplot
> # plot regressions on the given interval c(-0.06,0.06)
> plot(knn.reg.k1, add = TRUE, col="red",    xlim = c(-0.06,0.06))
> plot(knn.reg.k2, add = TRUE, col="green",  xlim = c(-0.06,0.06))
> plot(knn.reg.k3, add = TRUE, col="blue",   xlim = c(-0.06,0.06))
```

This code is used to produce Fig. 7.6. The function argument xis specifies the vector of regressors for the vector of dependent variables yis. The parameter k determines the number of neighbours with which to build the local average for the dependent variable. The argument x is a vector, which defines the interval for the regression analysis.

To achieve the best fit, one has to find the optimal k, similar to a kernel regression. It is not possible to establish a theoretical expression for the optimal value of k, since it depends greatly on the sample.

Fig. 7.6 Nonparametric regression of daily DAX log-returns by daily FTSE log-returns, using k-Nearest Neighbours. The plot shows fitted values for $k = 1$, $k = 10$ and $k = 250$.
Q BCS_kNN

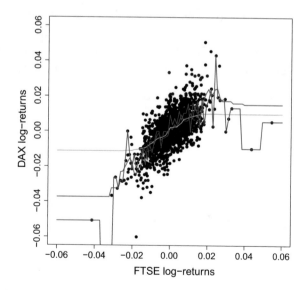

7.3.4 *Splines*

Another very famous nonparametric regression technique is the so-called smoothing spline. Here the method is different from those of the two previous nonparametric techniques. The spline regression is a general band-pass filter and is similar to the Hodrick–Prescott filter. This method does not look at how close the data is to the given point. Instead it imposes restrictions on the smoothness of the form of the regression. The smoothing is controlled by the parameter λ, which plays a similar role as the bandwidth h and the number of neighbours k.

The spline regression estimator $\hat{g}(x)$ is obtained by solving the following optimization problem:

$$\min_{g(x), g \in C^2} S_\lambda\{g(x)\} = \min_{g(x), g \in C^2} \sum_{i=1}^{n} \{y_i - g(x_i)\}^2 + \lambda \int \left[\frac{d^2 g(z)}{dz^2} \right]^2 dz. \quad (7.12)$$

The spline optimization problem (7.12) minimises the sum of the squared residuals and a penalty term. In most applications the penalty term is the second derivative of the estimator with respect to x, which reflects the smoothness of a function. The parameter λ determines the importance of the penalty term for the estimator. One can rewrite (7.12) in matrix notation since in fact the minimum in (7.12) is achieved by a piecewise cubic polynomial for $g(x)$:

$$S_\lambda\{g(x)\} = \{y - g(x)\}^\top \{y - g(x)\} + \lambda g(x)^\top K g(x). \quad (7.13)$$

It is possible to rewrite the sum $\sum_{i=1}^{n} \{y_i - g(x_i)\}^2$ as the inner product of the vector $y - g(x)$, where $y = \{y_1, \ldots, y_n\}^\top$ and the vector $g(x) = \{g(x_1), \ldots g(x_n)\}^\top$.

Then $g(z) = \sum_{i=1}^{n} g(x_i) p_i(z)$, where $p_i(z) = \sum_{k=0}^{3} a_{k,i} z^k$. The second derivative of $g(z)$ is $\frac{\partial^2 g(z)}{\partial z^2} = \sum_{i=1}^{n} g(x_i) \frac{\partial^2 p_i(z)}{\partial z^2} = g(x)^\top h(x)$, where $h(x) = \{ \frac{\partial^2 p(x)}{\partial x^2} \big|_{x=x_1}, \ldots, \frac{\partial^2 p(x)}{\partial x^2} \big|_{x=x_n} \}^\top$ is a vector containing the second derivatives of each of the cubic polynomials $p_i(x)$. The penalty term can be rewritten as follows:

$$\int \left[\frac{\partial^2 g(x)}{\partial x^2} \right]^2 dx = g(x)^\top h(x) h(x)^\top g(x) = g(x)^\top K g(x),$$

where the matrix $K_{n \times n}$ has entries $k_{i,j} = \int \frac{d^2 p_i(z)}{dz^2} \frac{d^2 p_j(z)}{dz^2} dz$.

Therefore the following estimator is proved to be a weighted sum of y:

$$\hat{g}(x) = \underset{g(x)}{\arg\min} S_\lambda\{g(x)\} = (\mathcal{I} + \lambda K)^{-1} y. \quad (7.14)$$

See Härdle et al. (2004) for a more detailed description. R provides in the package `stats` cubic splines for second derivative penalty terms using function `smooth.spline`.

Fig. 7.7 Nonparametric regression of daily DAX log-returns by daily FTSE log-returns, using spline regression. Regression results are depicted for $\lambda = 2$, $\lambda = 1$ and $\lambda = 0.2$. Ⓠ BCS_Splines

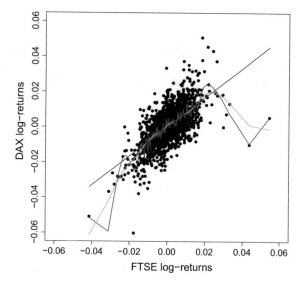

```
> r.dax    = diff(log(EuStockMarkets[, 1]))   # daily DAX log returns
> r.ftse   = diff(log(EuStockMarkets[, 4]))   # daily FTSE log returns
> # spline regressions
> sp1 = smooth.spline(x = r.ftse, y = r.dax, spar = 0.2)
> sp2 = smooth.spline(x = r.ftse, y = r.dax, spar = 1)
> sp3 = smooth.spline(x = r.ftse, y = r.dax, spar = 2)
> plot(r.dax, r.ftse)            # plot scatterplot
> lines(sp1, col ="red")         # plot regression line for span = 0.2
> lines(sp2, col ="green")       # plot regression line for span = 1
> lines(sp3, col ="blue")        # plot regression line for span = 2
```

This listing creates Fig. 7.7 for the regression of DAX log-returns and FTSE log-returns. The function arguments x and y are the observations of the independent and dependent variables, respectively. Instead of using two separate vectors, a matrix can be used. The argument spar defines λ through $\lambda = c^{3spar-1}$, therefore the greater the spar, the greater the λ. One can also apply weights to the observations of x through the variable w, which must have the same length as x. To find a good value for λ, set cv = TRUE for the ordinary cross-validation method and cv = FALSE for a generalised cross-validation method. Of course we could impose further restrictions on g, e.g. a penalty on its third derivative, or on any other type of norm.

7.3.5 LOESS or Local Regression

Another widespread method is the local regression, so-called LOESS or LOWESS. It is an improvement of the previous k-NN method, which aggregates the selection of neighbours of x within the $\{x_i\}$, adds a weighting to the sum of the y_i, and uses local polynomials for the fitting. For the univariate case, the model is

$$y_i = g(x_i) + \varepsilon_i.$$

The dependence between Y and X with samples $\{y_1, \ldots, y_n\}$ and $\{x_1, \ldots, x_n\}$ is approximated through a polynomial of order k evaluated at a focal point x. Therefore the polynomial regression function can be written as

$$y_i = \sum_{j=0}^{k} a_j(x_i - x)^j + \varepsilon_i.$$

In most applications, the observations are weighted according to their distance to x through the tri-cube weighting function

$$w(z) = \begin{cases} (1 - |z|^3)^3, & \text{if } |z| < 1, \\ 0, & \text{if } |z| \geq 1, \end{cases}$$

where $z_i = (x_i - x)/h$ and h is half the width of the interval around x. Therefore the weight attached to an observation is small if z is large and vice versa. This method has the same particularities as the k-NN method in the way that it can extrapolate the data. However, after the number of neighbours has been defined, through h, the coefficients a_j are estimated by the least squares approach.

This is both an advantage and a drawback, as the regression does not need any regularity conditions (for instance, compared to the spline method, which needs $\hat{g}$ to be twice differentiable), but it provides less intuition in the interpretation of the final curve (Fig. 7.8). The main function to use in R is loess(). First, one has to specify

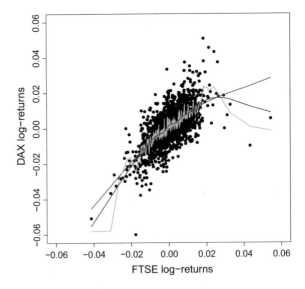

Fig. 7.8 Nonparametric regression of daily DAX log-returns by daily FTSE log-returns, using LOESS regression with degree one. The used LOESS parameters are: $\alpha = 0.9$, $\alpha = 0.3$ and $\alpha = 0.05$. ✪ BCS_LOESS

the two variables to regress in the syntax of linear regressions. Then the user needs to select the degree of the polynomial, and the `span` parameter, which represents the proportion of points (or neighbours) to use. By default, R uses the tri-cube weighting function $w(z)$, which the user can change. Nevertheless, the weight should satisfy some properties, stated in Cleveland (1979).

The following code produces a plot for a LOESS regression of DAX log-returns on FTSE log-returns.

```
> dax.r               = diff(log(EuStockMarkets[, 1]))
> ftse.r              = diff(log(EuStockMarkets[, 4]))
> loess1 = loess(r.dax ~ r.ftse,  # LOESS regression
+    degree = 1,                   # degree of polynomial
+    span   = 0.9)$fit             # proportion of neighbours
> loess2 = loess(r.dax ~ r.ftse, degree = 1, span = 0.01)$fit
> loess3 = loess(r.dax ~ r.ftse, degree = 1, span = 0.3)$fit
> l1 = loess1[order(r.ftse)]      # order as FTSE
> l2 = loess2[order(r.ftse)]
> l3 = loess3[order(r.ftse)]
> plot(x, y)
> lines(l1,col="red")
> lines(l2,col="green")
> lines(l3,col="blue")
```

This section and the previous sections introduced methods to model an unknown relation between two variables X and Y. Each method depends greatly on the smoothing parameter, which has different optimal values for different regression methods. As discussed in Sect. 5.1.4, the optimal bandwidth for a normal kernel is given by $h_{opt} = 1.06n^{-1/5}\hat{\sigma}$. The choice of the kernel becomes of minor importance as the number of observations increases. The optimal parameters for other methods are found via cross-validation algorithms.

On top of this, other complications can appear. Problems such as predicting from a low number of observations or the presence of outliers within the dataset can make the regression results less accurate. The following example illustrates, using simulated data, how different nonparametric regressions perform.

Example 7.1 Consider two rvs X and Y that are generated from the model

$$Y = g(X) + \varepsilon, \quad g(x) = \sin(2\pi x) - x^2, \quad X \sim N(0, 1) \ \ and \ \ \varepsilon \sim N(0, 1).$$

A small sample with $n = 50$ is simulated for this relation in R.

```
> # set.seed(3)                        # set seed, see Chap.\,9
> n                 = 50               # sample size
> Xis               = rnorm(n)         # random x
> Epsilon           = rnorm(n)         # random noise
> RegressionCurve = function(x) sin(2 * pi * x) - x^2
> Yis               = RegressionCurve(Xis) + Epsilon
```

One can use the rule of thumb for the kernel regression's bandwidth:

```
> kernel.reg.example = function(new.x){
+    ksmooth(x = Xis, y = Yis,
+       kernel     = "normal",
+       bandwidth  = 1.06 * n^(-1 / 5),
+       xpoints    = new.x)$y
+ }
```

However, for the k-NN regression and the spline regression, the smoothing parameter is selected by a cross-validation algorithm. Below is a simple line for the spline regression

```
> spline.reg.example = smooth.spline(x = Xis, y = Yis)
```

One can check that `spline.reg.example$spar` gives the smoothing parameter from a cross-validation algorithm.

A cross-validation algorithm has to be implemented to find the k-NN parameter k^{CV} optimal for the dataset at hand. In this example, the cross-validated parameter is

$$k^{CV} = \text{argmin}_k \, MSE(k) = \frac{1}{n}\sum_{i=1}^{n}\{\hat{y}_i(k) - y_i\}^2.$$

Therefore the value of k minimising $MSE(k)$ is used for the regression analysis. In the following, the leave-one-out cross-validation procedure is applied, where just one observation is dropped. Each observation will be excluded from the sample to preform the k-NN regression. Afterwards the squared error for each observation is computed for a specific k. The squared error is computed by the following code (Fig. 7.9).

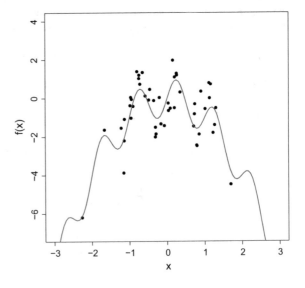

Fig. 7.9 Simulations along the regression curve $g(x) = \sin(2\pi x) - x^2$ displayed by points.
Q BCS_RegressionCurve

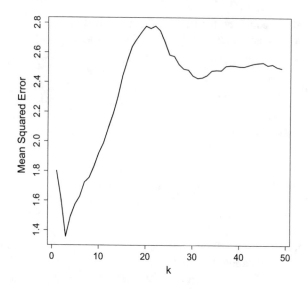

Fig. 7.10 MSE for *k*-NN regression using the Leave-one-out cross-validation method. ⧉ BCS_LeaveOneOut

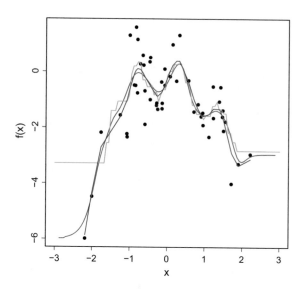

Fig. 7.11 Nonparametric regressions for simulated data. The regression results for kernel, kNN and spline are plotted. ⧉ BCS_NonparametricRegressions

```
> SEkNN = function(k, x, y, p){          # squared error function
+   yhat = knn.reg(x[p], x[-p], y[-p], k) # predicted y value
+   (yhat - y[p])^2                       # squared error
+ }
```

Now one needs to compute, for each k, the mean of the squared errors for every single point (x_i, y_i), and select the k that minimises the $MSE(k)$.

```
> listk  = matrix(0, n - 1, n)          # object for ks and es
> for (k in 1:(n - 1))                   # loop for possible ks
+    for (p in 1:n){                      # possible dropped obs.
+        listk[k, p] = SEkNN(k, Xis, Yis, p) # knn.reg required
+    }
> MSEkNN = (n)^(-1) * rowSums(listk)     # Mean squared error
> which.min(MSEkNN)                       # cross validated k
[1] 3
```

The code above shows that k^{CV} is equal to 3 for this dataset. Figure 7.10 plots the $MSE(k)$ depending on k for the k-NN regression. After the optimal parameters have been selected for each of the regressions, it is interesting to compare the results of these different methods to the true regression curve depicted in Fig. 7.9. Figure 7.11 shows that the regression curves are very similar. This actually tends to be true when n is large. In the meantime, all three regressions perform poorly at the boundaries of the support of x. The standard normally distributed rv X has 95% of its realisations in the interval $[-1.96, 1.96]$. Therefore the regression is likely to have both a large bias and a large variance outside of these bounds.

Chapter 8
Multivariate Statistical Analysis

Nothing in life is to be feared, it is only to be understood. Now is the time to understand more, so that we may fear less.

— Marie Curie

Multivariate procedures are at present widely used in finance, marketing, medicine and many other fields of theoretical and empirical research. This chapter introduces the basic tools of multivariate statistics in R.

The first part of the chapter deals with principal component analysis (PCA) and factor analysis (FA), which are methods for dimension reduction. Presented next is cluster analysis, also called data segmentation. It is a process of grouping objects into subsets, or 'clusters', such that objects within each cluster are more closely related to each other than to objects assigned to different clusters. Afterwards, multidimensional scaling is introduced. It is a statistical technique used in information visualisation to explore similarities and dissimilarities within data. Finally, discriminant analysis is discussed, which is a basic tool for linear classification.

8.1 Principal Components Analysis

One of the challenges of multivariate analysis, as discussed in the previous section, is the curse of dimensionality. The term refers to the problem, that the number of variables might be highly relative to the number of observations. This can be partly resolved by use of stepwise variable selection. But a high correlation between the original variables would lead to estimation and inference problems caused by near multicollinearity. This motivates principal component analysis (PCA), a multivariate technique whose central aim is to reduce the dimension of the dataset. This transformation leads to a new set of variables, which are linear combinations of the original variables, called principal components.

© Springer International Publishing AG 2017
W.K. Härdle et al., *Basic Elements of Computational Statistics*,
Statistics and Computing, DOI 10.1007/978-3-319-55336-8_8

There are several equivalent ways of deriving the principal components mathematically. The simplest way is to find the projections of the original p-dimensional vectors onto a subspace of dimension q. These projections should have the following property. The first component is the direction of the original variable space, along which the projection has the largest variance. The second principal component is the direction which maximises the variance among all directions orthogonal to the first principal component, and so on. Thus, the i-th component is the variance-maximising direction orthogonal to the previous $i - 1$ components. For an original dataset of dimension p, there are p principal components.

The principal component (PC) transformation of rv X with $\mathsf{E}(X) = \mu$ and $\mathsf{Var}(X) = \Sigma = \Gamma \Lambda \Gamma^\top$ is defined as

$$ Y = \Gamma^\top (X - \mu), $$

where Γ is the matrix of eigenvectors of the covariance matrix Σ and Λ is the diagonal matrix of the corresponding eigenvalues, see Sect. 2.1 for details. The principal component properties are given in the following theorem.

Theorem 8.1 *For a given* $X \sim (\mu, \Sigma)$*, let* $Y = \Gamma^\top (X - \mu)$ *be the principal component transformation. Then*

$$
\begin{aligned}
\mathsf{E}(Y_j) &= 0, & j &= 1, \ldots, p; \\
\mathsf{Var}(Y_j) &= \lambda_j, & j &= 1, \ldots, p; \\
\mathsf{Cov}(Y_i, Y_j) &= 0, & i &\neq j; \\
\mathsf{Var}(Y_1) &\geq \mathsf{Var}(Y_2) \geq \ldots \geq \mathsf{Var}(Y_p) > 0. &&
\end{aligned}
$$

In practice, the expectation μ and covariance matrix Σ are replaced by their estimators $\bar{x}$ and S respectively. If $S = \mathcal{G}\mathcal{L}\mathcal{G}^\top$ is the spectral decomposition of S, then the principal components are obtained by $\mathcal{Y} = (\mathcal{X} - 1_n \bar{x}^\top)\mathcal{G}$. Note that with the centring matrix $\mathcal{H} = \mathcal{I} - n^{-1}1_n 1_n^\top$ and $\mathcal{H} 1_n \bar{x}^\top = 0$, the empirical covariance matrix of the principal components can be written as

$$ S_y = n^{-1}\mathcal{Y}^\top \mathcal{H}\mathcal{Y} = \mathcal{L}, $$

where $\mathcal{L} = \text{diag}(l_1, \ldots, l_p)$ is the matrix of eigenvalues of S. Details on principal components theory and their properties are given in Härdle and Simar (2015).

It is important to clarify the meaning of the components. In some cases, the components actually measure real variables, while in others they just reflect patterns of the variance-covariance matrix.

Let us illustrate PCA with an example on genuine and counterfeit banknotes, provided by Riedwyl (1997) and included in the package `mclust`. The data set contains seven variables and 200 observations. The first variable is coded 0/1 and states whether the banknote is genuine or not. For this analysis, only the last 6 variables are used. These are quantitative characteristics of swiss banknotes (e.g.

length, width, etc.). First, the required package is loaded and the data set is saved as a data frame without the indicator stating whether banknotes are genuine.

```
> data(banknote, package = "mclust")   # load the data
> mydata = banknote[, -1]              # remove the first column
```

To perform the principal component analysis, there exists a built-in function `princomp`.

```
> fit = princomp(mydata)              # compute PCA
> summary(fit)                        # print results w.o. loadings
Importance of components:
                        Comp.1 Comp.2 Comp.3 Comp.4 Comp.5 Comp.6
Standard deviation        1.73   0.96  0.492  0.440  0.291 0.1880
Proportion of Variance    0.67   0.21  0.054  0.043  0.019 0.0079
Cumulative Proportion     0.67   0.88  0.930  0.973  0.992 1.0000
```

The output includes the standard deviations of each component, i.e. the square root of the covariance matrix's eigenvalues. A measure of how well the first q PCs explain the total variance is given by the cumulative relative proportion of variance

$$\psi_q = \frac{\sum_{j=1}^{q} \lambda_j}{\sum_{j=1}^{p} \lambda_j}.$$

The loadings matrix, given by matrix Γ, gives the multiplicative weights of each standardised variable in the component score. In practice, one considers the matrix $\mathcal{G}$. Small loadings values are replaced by a space in order to highlight the pattern of loadings.

```
> print(fit$loadings, digits = 3)              # prints loadings

Loadings:
          Comp.1 Comp.2 Comp.3 Comp.4 Comp.5 Comp.6
Length                  -0.326  0.562  0.753
Left       0.112        -0.259  0.455 -0.347 -0.767
Right      0.139        -0.345  0.415 -0.535  0.632
Bottom     0.768 -0.563 -0.218 -0.186
Top        0.202  0.659 -0.557 -0.451  0.102
Diagonal  -0.579 -0.489 -0.592 -0.258


              Comp.1 Comp.2 Comp.3 Comp.4 Comp.5 Comp.6
SS loadings     1.00   1.00   1.00   1.00   1.00   1.00
Proportion Var  0.17   0.17   0.17   0.17   0.17   0.17
Cumulative Var  0.17   0.33   0.50   0.67   0.83   1.00
```

Scatter plots in Fig. 8.1 of the components clearly illustrate how principal component analysis simplifies multivariate techniques.

```
> layout(matrix(1:4, 2, 2))
> group = factor(banknote[, 1])               # group as factor
> plot(fit$scores[, 1:2], col = group)        # plot 1 vs 2 factor
> plot(fit$scores[, c(1, 3)], col = group)    # plot 1 vs 3 factor
> plot(fit$scores[, 2:3], col = group)        # plot 2 vs 3 factor
```

In practice, a question which often arises is how to choose the number of components. Commonly, one retains just those components that explain some specified percentage

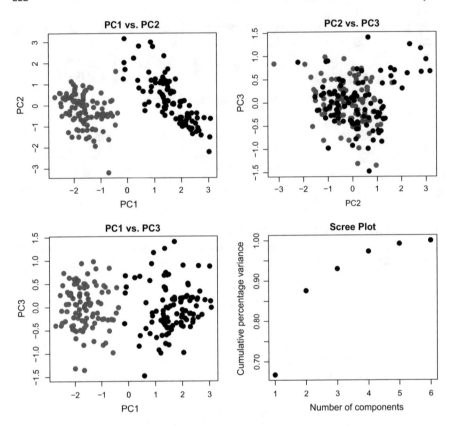

Fig. 8.1 Principal components and scree plot for Swiss Banknote dataset. ⌀ BCS_PCAvar

of the total variation of the original variables. Values between 70 and 90% are usually suggested, although smaller values might be appropriate as p or the sample size increase.

```
> plot(cumsum(fit$sdev^2 / sum(fit$sdev^2)))     # cumulative variance
```

A graphical representation of the PCs' ability to explain the variation in the data is given in Fig. 8.1. The plot down on the right, called *scree plot*, depicts the relative cumulative proportion of the explained variance as given by ψ_q above. The figure implies that the use of the first and the second principal components is sufficient to identify the genuine banknotes.

Another way to choose the optimal number of principal components is to exclude the principal components with eigenvalues less than the average, see Everitt (2005).

The covariance between the PC vector Y and the original variables X is important for the interpretation of the PCs. It is calculated as

$$\text{Cov}(X, Y) = \Gamma \Lambda.$$

Hence, the correlation $\rho_{X_i Y_j}$ between variable X_i and the PC Y_j is

$$\rho_{X_i Y_j} = \frac{\gamma_{ij} \lambda_j}{(\sigma_{X_i X_i} \lambda_j)^{1/2}} = \gamma_{ij} \left(\frac{\lambda_j}{\sigma_{X_i X_i}} \right)^{1/2},$$

where γ_{ij} is the eigenvector corresponding to the eigenvalue λ_j.

In practice, all variances, eigenvectors and eigenvalues are replaced by their estimators to calculate the empirical correlation $r_{X_i Y_j}$. Note that $\sum_{j=1}^{p} r_{X_i Y_j}^2 = 1$.

```
> corr = cor(mydata, fit$scores)   # correlation of PC and variables
> cev  = cbind(corr[, 1:2],        # cumulative
+              corr[, 1]^2 + corr[, 2]^2)
> print(cev, digits = 2)           # correlations and communalities
           Comp.1 Comp.2
Length      -0.20  0.028 0.041
Left         0.54  0.191 0.326
Right        0.60  0.159 0.381
Bottom       0.92 -0.377 0.991
Top          0.44  0.794 0.820
Diagonal    -0.87 -0.410 0.925
```

The correlations of the original variables X_i with the first two PCs are given in the first two columns of the table in the previous code. The third column shows the cumulative percentage of the variance of each variable explained by the first two principal components Y_1 and Y_2, i.e. $\sum_{j=1}^{2} r_{X_i Y_j}^2$.

The results are displayed visually in a correlation plot, where $r_{X_i Y_1}^2$ are plotted against $r_{X_i Y_2}^2$ in Fig. 8.2 (left). When the variables lie near the periphery of the circle, they are well explained by the first two PCs. The plot confirms that the percentage of the variance of X_1 explained by the first two PCs is relatively small.

```
> # coordinates for the surrounding circle
> ucircle = cbind(cos((0:360) / 180*pi), sin((0:360) / 180*pi))
> plot(ucircle, type = "l", lty = "solid")   # plot circle
> abline(h = 0.0, v = 0.0)                    # plot orthogonal lines
> label = paste("X", 1:6, sep = "")
> text(cor(mydata, fit$scores), label)        # plot scores in text
```

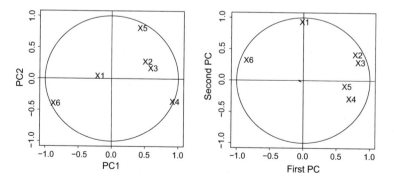

Fig. 8.2 The correlation of the original variable with the PCs *(left)* and normalised PCs *(right)*.
Q BCS_PCAbiplot, Q BCS_NPCAbiplot

The PC technique is sensitive to scale changes. If a variable is multiplied by a scalar, different eigenvalues and eigenvectors are obtained. This is because the eigenvalue decomposition is performed on the empirical covariance matrix S and not on the empirical correlation matrix $\mathcal{R}$. For this reason, in certain situations, variables should be standardised by the Mahalanobis transformation

$$\mathcal{X}' = \mathcal{H}\mathcal{X}\mathcal{D}^{-1/2},$$

where $\mathcal{D} = \mathrm{diag}(s_{X_1X_1}, \ldots, s_{X_pX_p})$ with covariances $s_{X_1X_1}, \ldots, s_{X_pX_p}$. Due to this standardisation, the means of the transformed variables $\overline{x'}_1 = \ldots = \overline{x'}_p = 0$ and the new variance-covariance matrix is $S' = \mathcal{R}$, see Theorem 6.2. The PCs obtained from the standardised matrix are usually called normalised principal components. The transformation is done to avoid heterogeneity in the variables with respect to their covariances, which can be the case when variables are measured on heterogeneous scales (e.g. years, kilograms, dollars). To perform the normalised principal component analysis, the former R code is changed to

```
> princomp(mydata, cor = TRUE)    # calculate normalised PCs
```

The scree plot for the normalised model would be different and the variables can be observed to lie closer to the periphery of the circle in Fig. 8.2 (right).

8.2   Factor Analysis

Factor analysis is widely used in behavioural sciences. Scientists are often interested not in the observed variables, but in unobserved factors. For example, sociologists record people's occupation, education, home ownership, etc., on the assumption that these factors reflect their unobservable 'social class'. Explanatory factor analysis investigates the relationship between manifested variables and factors. Note that the number of variables should be much smaller than the number of observations. The factor model can be written as

$$X = QF + \mu, \tag{8.1}$$

where X is a $p \times 1$ vector of observable random variables, μ is the mean vector of X, F is a $k \times 1$-dimensional vector of *(unobservable) factors* and Q is a $p \times k$ *matrix of loadings*. It is also assumed that $\mathsf{E}(F) = 0$ and $\mathsf{Var}(F) = \mathcal{I}$.

In practice, factors are usually split into specific factors and common factors, which are highly informative and common to all of the components of X. In other words, the factor model explains the variation of X by a small number of latent factors F, which are common to the p components of X, and an *individual* factor U, which allows for component-specific variation. Thus, a generalisation of (8.1) is given by

$$X = QF + \mu + U,$$

where U is a $(p \times 1)$ vector of the specific factors, which are assumed to be random. Additionally, it is assumed that $\mathsf{Cov}(U, F) = 0$ and $\mathsf{Cov}(U_j, U_k) = 0$, for $j \neq k \in \{1, \ldots, p\}$. The covariance matrix of X can then be written as

$$\Sigma = QQ^\top + \psi, \tag{8.2}$$

where $QQ^\top$ is called the *communality* and ψ is the *specific variance*.

To interpret a specific factor F_j, its correlation with the original variables is computed. The covariance between X and F is given by

$$\Sigma_{XF} = \mathsf{E}\{(QF + U)F^\top\} = Q.$$

The correlation is

$$\mathcal{P}_{XF} = \mathcal{D}^{-1/2} Q,$$

where $\mathcal{D} = \mathrm{diag}(\sigma_{X_1 X_1}, \ldots, \sigma_{X_p X_p})$.

The analysis based on the normalised variables is performed by using $\mathcal{R} = QQ^\top + \psi$ and the loadings can be interpreted directly. Factor analysis is scale invariant. However, the loadings are unique only up to multiplication by an orthogonal matrix. This leads to potential difficulties with estimation, but facilitates the interpretation of the factors. Multiplication by an orthogonal matrix is called rotation of the factors. The most widely used rotation is the varimax rotation which maximises the sum of the variances of the squared loadings within each column. For more details on factor analysis see Rencher (2002).

In practice, Q and ψ have to be estimated by $S = \hat{Q}\hat{Q}^\top + \hat{\psi}$. The number of estimated parameters is $d = \frac{1}{2}(p-k)^2 - \frac{1}{2}(p+k)$. An exact solution exists only when $d = 0$, otherwise an approximation must be used. Assuming a normal distribution of the factors, maximum likelihood estimators can be computed as discussed below. Other methods to find $\hat{Q}$ are principal factors and principal component analysis, see Härdle and Simar (2015) for details.

8.2.1 Maximum Likelihood Factor Analysis

This subsection takes a closer look at the maximum likelihood factor analysis, one possible fitting procedure in factor analysis. It is generally recommended if it can be assumed that the factor scores are independent across factors and individuals and normally distributed, i.e. $F \sim N(0, 1)$. Under this assumption, it is necessary that $X \sim N(0, \Psi + w^\top w)$ and $\widehat{\Psi}$ and $\hat{w}$ are found by maximising the log-likelihood

$$L = -\frac{np}{2} \log 2\pi - \frac{n}{2} \log |\Psi + w^\top w| - \frac{n}{2} \mathrm{tr}\left\{(\Psi + w^\top w)^{-1} V\right\}.$$

Maximum likelihood factor analysis is implemented in R in the function `factanal()`. Its arguments are the number of factors and the factor rotation method. It will be applied to a famous example, the analysis of the performance of decathlon athletes, see Everitt and Hothorn (2011) and Park and Zatsiorsky (2011). Studies of this type are motivated by the desire to determine the overall success in a decathlon and to help instructors and athletes with the design of optimal training programs. They therefore need to consider the inter-event similarity and possible transfer of training results. The data is part of the package `FactoMineR` and consists of 41 rows and 13 columns. The first ten columns correspond to the performance of the athletes in the 10 events of the decathlon. Columns 11 and 12 correspond to the rank and the points obtained respectively. The last column is a categorical variable corresponding to the sporting event (2004 Olympic Games or 2004 Decastar).

In order to choose a reasonable number of factors, cumulative explained variance is plotted against number of principal components beforehand.

```
> require(stats)
> data("decathlon", package = "FactoMineR")
> mydata = decathlon[, 1:10] # choose relevant variables
> fit    = princomp(mydata)  # perform PCA
> # plot cum. percentage of var. explained by number of components
> plot(cumsum(fit$sdev^2 / sum(fit$sdev^2)),
+     xlab = "Number of principal components",
+     ylab = "Cumulative percentage variance")
```

Figure 8.3 suggests to start the analysis with three factors, because the increase in explained variance becomes very small for more than three factors. While including a third factor increases the explained variance by about 6 percentage points, including a fourth factor offers an increase of less than one additional percentage point.

```
> require(stats)
```

Fig. 8.3 The correlation of the original variable with the NPCs. **Q** BCS_FAsummary

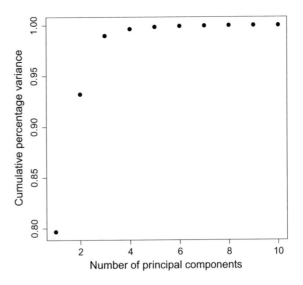

```
> mydata  =  decathlon[,  1:10]
> fit      =  factanal(mydata,      # fit the model
+    factors   = 3,                  # number of factors
+    rotation = "none")              # no rotation performed
> fit                               # print the results

Call:
factanal(x  =  mydata,  factors  =  3,  rotation  = "none")

Uniquenesses:
      100m           Long.jump  Shot.put     High.jump   400m
      0.411          0.396      0.106        0.697       0.264
      110m.hurdle    Discus     Pole.vault   Javeline    1500m
      0.491          0.534      0.907        0.785       0.005
```

The output states the computed uniquenesses and a matrix of loadings with one column for each factor. Uniqueness gives the proportion of variance of the variable not associated with the factors. It is defined as $1 - $ *communality* (namely $1 - \sum_{l=1}^{k} q_{jl}^2$, $j = 1, \ldots, k$), where communality is the variance of that variable as determined by the common factors. Note that the greater the uniqueness of a variable, the lower the relevance of the variable in the factor model, since the factors capture less of the variance of the variable.

```
Loadings:                             # output from fit, cont.
                Factor1 Factor2 Factor3
100m            -0.573   0.507
Long.jump        0.455  -0.629
Shot.put         0.881   0.322   0.121
High.jump        0.547
400m            -0.432   0.617   0.412
110m.hurdle     -0.493   0.514
Discus           0.621   0.114   0.261
Pole.vault              -0.174   0.246
Javeline         0.356   0.239  -0.178
1500m                            0.997


                Factor1 Factor2 Factor3
SS loadings       2.554   1.504   1.347
Proportion Var    0.255   0.150   0.135
Cumulative Var    0.255   0.406   0.541
```

```
Test of the hypothesis that 3 factors are sufficient.
The chi square statistic is 17.97 on 18 degrees of freedom.
The p-value is 0.457
```

The factor loadings give the correlation between the factors and the observed variables. They can be used to interpret the factors based on the variables they capture. As stated above, the factor loadings are not unique to multiplication by an orthogonal matrix and this multiplication, called *rotation* of the factor loadings matrix, can greatly facilitate interpretation.

The most common rotation method, *varimax*, aims at maximising the variance of the squared loadings of a factor on all the variables. Note that the assumption of orthogonality of the factors is required. There is a number of 'oblique' rotations available, which allow the factors to correlate.

```
> mydata  =  decathlon[,  1:10]
```

```
> fit2    = factanal(mydata,    # factor analysis
+    factors  = 3,               # 3 factors
+    rotation = "varimax")       # varimax rotation
> fit2                           # print the results

Call:
factanal(x = mydata, factors = 3, rotation = "varimax")
```

```
Uniquenesses:
        100m            Long.jump Shot.put   High.jump  400m
        0.411           0.396     0.106      0.697      0.264
        110m.hurdle     Discus    Pole.vault Javeline   1500m
        0.491           0.534     0.907      0.785      0.005
```

```
Loadings:
            Factor1 Factor2 Factor3
100m          0.699  -0.264  -0.178
Long.jump    -0.764           0.107
Shot.put     -0.128   0.934
High.jump    -0.232   0.497
400m          0.815           0.263
110m.hurdle   0.685  -0.183
Discus       -0.152   0.618   0.248
Pole.vault   -0.124           0.278
Javeline               0.412  -0.214
1500m         0.190           0.976
```

```
                Factor1 Factor2 Factor3
SS loadings       2.350   1.791   1.264
Proportion Var    0.235   0.179   0.126
Cumulative Var    0.235   0.414   0.541
```

```
Test of the hypothesis that 3 factors are sufficient.
The chi square statistic is 17.97 on 18 degrees of freedom.
The p-value is 0.457
```

For the first factor, the largest loadings are for 100 m, 400 m, 110 m hurdle run, and long jump. This factor can be interpreted as the 'sprinting performance'. The loadings for the second factor present a counter-intuitive throwing-jumping combination: the highest loadings are for the three throwing events (discus throwing, javeline and shot put) and for the high jump event. For the third factor, the largest loading is for 1500 m running. The first and second factors can be interpreted straightforward as 'sprinting abilities' and 'endurance'. The meaning of the last factor is not evident.

Note that the model fit has room for improvement, because the value 0.54 for Cumulative Var in the third line signifies that only 54% of the variation in the data is explained by three factors. Including a fourth factor is easily done by changing the code to factors = 4.

```
> mydata = decathlon[, 1:10]
> fit3    = factanal(mydata,    # factor model
+    factors  = 4,               # 4 factors
+    rotation = "varimax")       # varimax rotation
> fit3                           # print the results

Call:
factanal(x = mydata, factors = 4, rotation = "varimax")
```

```
Uniquenesses:
         100m          Long.jump  Shot.put    High.jump   400m
         0.409         0.386      0.005       0.680       0.270
         110m.hurdle   Discus     Pole.vault  Javeline    1500m
         0.464         0.492      0.005       0.800       0.005

Loadings:
              Factor1 Factor2 Factor3 Factor4
100m           0.720  -0.245  -0.112
Long.jump     -0.770                   0.131
Shot.put      -0.144   0.976   0.103   0.103
High.jump     -0.259   0.480          -0.152
400m           0.770           0.363
110m.hurdle    0.712  -0.157
Discus        -0.220   0.585   0.297  -0.170
Pole.vault    -0.102           0.117   0.983
Javeline               0.403  -0.191
1500m                          0.984   0.143

              Factor1 Factor2 Factor3 Factor4
SS loadings     2.363   1.785   1.263   1.074
Proportion Var  0.236   0.178   0.126   0.107
Cumulative Var  0.236   0.415   0.541   0.648

Test of the hypothesis that 4 factors are sufficient.
The chi square statistic is 9.2 on 11 degrees of freedom.
The p-value is 0.603
```

This improves the result, with the new model explaining 65% of variation. The interpretation of the first and second factor remains the same, but the 1500 m run and pole vaulting are now captured by factors 3 and 4, respectively.

Thus, there is no unique or 'best' solution in factor analysis. Using the maximum likelihood method allows to test the goodness of the factor model. The test examines if the model fits significantly worse than a model in which the variables correlate freely. p-values higher than 0.05 indicate a good fit, since the null hypothesis of a good fit cannot be rejected.

In this case, the p-value is 0.457. The null hypothesis that 3 factors are sufficient cannot be rejected, suggesting a good fit of the model.

8.3 Cluster Analysis

Cluster analysis techniques are used to search for clusters or groups in a priori unclassified multivariate data. The main goal is to obtain clusters of objects which are similar to one another and different from objects in other clusters. Many methods of cluster analysis have been developed, since most studies allow for a variety of techniques.

8.3.1 Proximity of Objects

The starting point for cluster analysis is a $(n \times p)$ data matrix $\mathcal{X}$ with n measurements of p objects. The proximity among objects is described by a matrix $\mathcal{D}$ which contains measures of similarity or dissimilarity among the n objects. The elements can be either distance or proximity measures. The nature of the observations plays an important role in the choice of the measure. Nominal values lead, in general, to proximity values, whereas metric values lead to distance matrices. To measure the similarity of objects with binary structure, one defines

$$a_{ij,1} = \sum_{k=1}^{p} I\,(x_{ik} = x_{jk} = 1), \qquad a_{ij,2} = \sum_{k=1}^{p} I\,(x_{ik} = 0, x_{jk} = 1),$$

$$a_{ij,3} = \sum_{k=1}^{p} I\,(x_{ik} = 1, x_{jk} = 0), \qquad a_{ij,4} = \sum_{k=1}^{p} I\,(x_{ik} = x_{jk} = 0).$$

The following proximity measure is used in practice:

$$d_{ij} = \frac{a_{ij,1} + \delta a_{ij,4}}{a_{ij,1} + \delta a_{ij,4} + \lambda(a_{ij,2} + a_{ij,3})},$$

where δ and λ are weighting factors. Table 8.1 shows two similarity measures for given weighting factors. To measure the distance between continuous variables, one uses L_r-norms (see Sect. 2.1.5):

$$d_{ij} = ||x_i - x_j||_r = \left\{ \sum_{k=1}^{p} |x_{ik} - x_{jk}|^r \right\}^{1/r}, \tag{8.3}$$

where x_{ik} denotes the value of the k-th variable of object i. The class of distances in (8.3) measures the dissimilarity using different weights for varying r. The L_1-norm, for example, gives less weight to outliers than the L_2-norm (the Euclidean norm).

An underlying assumption in applying L_r-norms see Sect. 2.1.5 is that the variables are measured on the same scale. Otherwise, a standardisation is required, corresponding to a more general L_2 or Euclidean norm with a matrix $\mathcal{A}$, where $\mathcal{A} > 0$:

$$d_{ij}^2 = ||x_i - x_j||_{\mathcal{A}}^2 = (x_i - x_j)^{\top} \mathcal{A}(x_i - x_j).$$

Table 8.1 Common similarity coefficients

Name	δ	λ	Definition
Jaccard	0	1	$\dfrac{a_1}{a_1 + a_2 + a_3}$
Tanimoto	1	2	$\dfrac{a_1 + a_4}{a_1 + 2(a_2 + a_3) + a_4}$

L_2-norms are given by $\mathcal{A} = \mathcal{I}_p$, but if a standardisation is desired, then the weight matrix $\mathcal{A} = \mathrm{diag}(s_{X_1 X_1}^{-1}, \ldots, s_{X_p X_p}^{-1})$ is suitable. Recall that $s_{X_k X_k}$ is the empirical variance of the k-th component, hence

$$d_{ij}^2 = \sum_{k=1}^{p} \frac{(x_{ik} - x_{jk})^2}{s_{X_k X_k}}.$$

Here each component has the same weight and the distance does not depend on any particular measurement units.

In practice, the data is often mixed, i.e. contains both binary and continuous variables. One way to solve this is to recode the data to normalised similarity by assigning each attribute level to a separate binary variable. However, this approach often does not adequately capture the size of distance that can be captured by continuous variables and leads to a large increase in a_4.

The second way is to calculate a generalised similarity measure, e.g. the commonly used Gower similarity coefficient. It calculates and average over similarities and is defined as

$$D_{ij} = \frac{\sum_{k=1}^{v} d_{ijk} \cdot \delta_{ijk}}{\delta_{ijk}},$$

where D_{ij} is the Gower proximity, δ_{ijk} is 0 if $x_i = x_j = 0$ and 1 else, d_{ijk} is the similarity between objects i and j of attribute k. d_{ijk} is defined differently for different types of variables:

- For binary dichotomous variables, $d_{ijk}=1$ if $x_{ik} = x_{jk} = 1$, similar to a_1 above.
- For binary qualitative variables, $d_{ijk}=1$ if $x_{ik} = x_{jk} = 1$ or $x_{ik} = x_{jk} = 0$, similar to a_1 and a_2 above.
- For continuous variables, $d_{ijk} = 1 - \frac{|x_{ik} - x_{jk}|}{R_k}$, where R_k is the range of attribute k in the sample (or population).

Note that the Gower proximity coefficient reduces to the Jaccard similarity index if all attributes are binary. If all attributes are qualitative then it reduces to the *Simple Matching* coefficient. When all variables are quantitative (interval) then the coefficient is the range-normalised *City-block metric*.

8.3.2 Clustering Algorithms

There are two types of hierarchical clustering algorithms, agglomerative and splitting. The first starts with the finest possible partition. The second starts with the coarsest possible partition, i.e. one cluster containing all the observations. The drawback of these methods is that the clusters are not adjusted, e.g. objects assigned to a cluster cannot be removed in further steps. Since all agglomerative hierarchical techniques ultimately reduce the data to a single cluster containing all the individuals, the investigator seeking the solution with the best-fitting number of clusters will need to decide which division to choose. The problem of deciding on the 'correct' number of clusters will be taken up later.

The hierarchical agglomerative clustering algorithm works as follows:

1. Find the nearest pair of distinct clusters, say c_i and c_j, merge them into c_k and decrease the number of clusters by one;
2. If the number of clusters equals one, end the algorithm, else return to step 1.

For this purpose, the distance between two groups or an individual and a group must be calculated. Different definitions of distances lead to different clustering algorithms. Widely used measures of distance are

$$\text{Single linkage}\quad d_{AB} = \min_{i \in A,\, j \in B} d_{ij},$$

$$\text{Complete linkage}\quad d_{AB} = \max_{i \in A,\, j \in B} d_{ij},$$

$$\text{Average linkage}\quad d_{AB} = \frac{1}{n_A n_B} \sum_{i \in A} \sum_{j \in B} d_{ij},$$

where n_A and n_B are the number of objects in the two groups. The Ward clustering algorithm does not put together groups with smaller distance. Instead, it joins groups that do not increase a given measure of heterogeneity 'too much'. The resulting groups are as homogeneous as possible. Let us study an example of the Gross National Product (GNP) per capita and the percentage of the population working in agriculture for each country belonging to the European Union in 1993. The data can be loaded from the cluster package. First the data should be checked for missing values and standardised. As mentioned above, the matrix of distances should be calculated first. Euclidean distance is used in this case. This matrix is then used to obtain clusters using the complete linkage algorithm (other algorithms are available as parameters of function hclust).

```
> require(cluster)                              # package for CA
> data(agriculture, package ="cluster")         # load the data
> mydata = scale(agriculture)                   # normalise data
> d       = dist(mydata,                        # calculate distances
+      method ="euclidean")                      # Euclidean
> print(d, digits = 2)                          # show distances
         B     DK     D    GR     E     F   IRL     I     L    NL     P
DK    1.02
D     0.40  0.63
GR    3.74  4.03  3.88
E     1.68  2.15  1.87  2.08
F     0.55  0.71  0.43  3.48  1.50
IRL   2.12  2.46  2.27  1.62  0.49  1.87
I     0.90  1.04  0.88  3.03  1.11  0.46  1.43
L     0.86  0.35  0.46  4.21  2.25  0.75  2.61  1.18
NL    0.26  1.01  0.48  3.49  1.44  0.39  1.87  0.65  0.94
P     2.92  3.27  3.08  0.84  1.24  2.68  0.82  2.25  3.43  2.67
UK    0.57  1.56  0.97  3.49  1.43  0.96  1.92  1.10  1.42  0.58  2.66
> fit = hclust(d, method ="complete")           # fit the model
```

Fig. 8.4 Agglomerative
hierarchical clustering for
agriculture data.
Q BCS_CAComplete

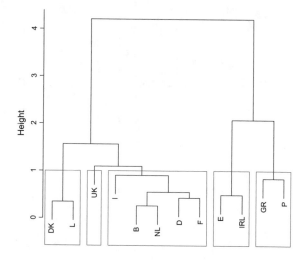

The specific partition of the data can now be selected from the dendrogram, see
Fig. 8.4. 'Cutting' off the dendrogram at some height will give a partition with a
particular number of groups. One of the methods to choose the number of clusters
is to examine the size of the height changes in the dendrogram, where a large jump
in the dendrogram indicates a large loss in homogeneity of clusters if they are joint
as suggested by the current step of the tree. Function `rect.hclust` draws a dendrogram with red borders around clusters, facilitating the interpretation, see Fig. 8.4
where axis `height` shows the value of the criterion associated with the clustering
method. The value `k` specifies the desired number of groups. We do not discuss the
choice of the number of clusters in detail. A popular method is the scree plot or elbow
criterion, which is easy to implement and visualise.

```
> plot(fit)                             # plot the solution
> groups = cutree(fit, k = 5)           # define clusters
> rect.hclust(fit, k = 5, border = "red")   # draw boxes
```

The importance of the method choice is illustrated in Fig. 8.5.

```
> par(mfrow = c(1, 3))                                    # 3 plots in 1 figure
> plot(hclust(d, method = "single"),  main = "Single linkage")
> plot(hclust(d, method = "ward.D"),  main = "Ward")
> plot(hclust(d, method = "average"), main = "Average linkage")
```

If the number of clusters is predetermined by the application, *k-means clustering* can
be used to partition the observations into a pre-specified number k of clusters. The
clusters $S_1, S_2, \ldots, S_k$ are chosen so as to minimise the Euclidean distance between
the points and the mean within each cluster, also called the *within-cluster sum of
squares*. This can be expressed as

$$\arg\min_{S} \sum_{i=1}^{k} \sum_{x \in S_i} \|x - \mu_i\|^2 ,$$

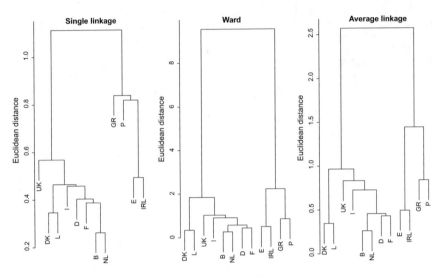

Fig. 8.5 Comparison of different clustering algorithms. ▢ BCS_CAmethods

where μ_i is the centroid in S_i. This is implemented in function `kmeans()`, in which option `centres` specifies the number of clusters.

Cluster analysis is a large area of multidimensional statistical analysis and has been covered only very briefly in this section. For a detailed discussion of this technique see Everitt et al. (2009).

8.4 Multidimensional Scaling

Multidimensional scaling (MDS) is an exploratory technique used to visualise proximities in a low dimensional space. Interpretation of the dimensions can lead to an understanding of the processes underlying the perceived nearness of objects. Furthermore, it is possible to incorporate individual or group differences in the solution. The basic data representation in a standard MDS is a dissimilarity matrix that shows the distance between every possible pair of objects. Using spectral decomposition of this matrix, the desired projections of the data to the lower dimension are found. The goal of MDS is to faithfully represent these distances with the lowest possible dimensional space.

The variety of methods that have been proposed largely differ in how agreement between fitted distances and observed proximities is assessed. In this section, classical metric MDS and non-metric MDS are considered. Metric MDS is applied when measurements are numerical and the distance between the objects can be calculated. In non-metric MDS, proximity measurements are ordinal, e.g. when a subject prefers

Coca-Cola to Pepsi-Cola, but cannot say how much. This kind of data is used very often in psychology and market research.

8.4.1 Metric Multidimensional Scaling

Assume that a data matrix $\mathcal{X}$ is given. The metric MDS begins with a $n \times n$ distance matrix $\mathcal{D}^{\mathcal{X}}$, which contains distances between the given objects. Note that this matrix is symmetric, with $d_{ii}^{\mathcal{X}} = 0$ and $d_{ij}^{\mathcal{X}} > 0$, which naturally follows from the definition of distance in any metric space. Given such a matrix, MDS attempts to find n data points $y_1, \ldots, y_n$ constituting the new data matrix $\mathcal{Y}$ in p-dimensional space, such that $\mathcal{D}^{\mathcal{X}}$ is similar to $\mathcal{D}^{\mathcal{Y}}$. In particular, metric MDS minimises

$$\min_{y} \sum_{i=1}^{n} \sum_{j=1}^{n} (d_{ij}^{\mathcal{X}} - d_{ij}^{\mathcal{Y}})^2, \tag{8.4}$$

where $d_{ij}^{\mathcal{X}} = \|x_i - x_j\|$ and $d_{ij}^{\mathcal{Y}} = \|y_i - y_j\|$. For the Euclidean distance, (8.4) can be reduced to

$$\min_{y} \sum_{i=1}^{n} \sum_{j=1}^{n} (x_i^\top x_j - y_i^\top y_i)^2.$$

In practice, the question how to choose the dimension of data projection p arises often. If one defines $P = \sum_{i=1}^{p} |\lambda_i| / \sum_{i=1}^{n} |\lambda_i|$ and $P' = \sum_{i=1}^{p} \lambda_i^2 / \sum_{i=1}^{n} \lambda_i^2$, then a p which gives $P > 0.8$ or $P' > 0.8$ suggests a reasonable fit. The λ_i's are the first p eigenvalues of the data matrix $\mathcal{X}$. These values do not necessary agree. Usually p is chosen such that the conditions hold for P and P' simultaneously. Another criterion is to choose p such that the sum of the largest positive eigenvalues is approximately equal to the sum of all eigenvalues. A third criterion proposes to accept as genuinely positive only those eigenvalues whose magnitude substantially exceeds that of the largest negative eigenvalue.

MDS is widely used in psychological sciences and marketing research. The Cars93 data frame (package MASS) has 93 rows and 27 variables, which are displayed by the command View(Cars93). Only models with rear drive train and with less than 18 mpg (miles per gallon) in the city are analysed and only numerical characteristics taken into account. First, the data is prepared for analysis.

```
> data(Cars93, package = "MASS")                        # load the data
> rownames(Cars93)  = Cars93[, ncol(Cars93)]
> mydata = Cars93[which(Cars93$DriveTrain == "Rear"      # choose
+     & Cars93$MPG.city <= 18),                          # cars and
+    c(5, 7:8, 11:15, 17:19, 20:25)]                     # variables
> mydata = na.omit(mydata)                               # exclude missing
> d      = dist(mydata, method = "euclidean")            # distance matrix
```

Fig. 8.6 Configuration of the MDS of the American car subsample. **Q** BCS_MDS

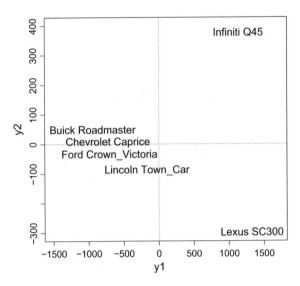

Function `cmdscale()` performs MDS and uses as arguments the distance matrix and the dimension of the space in which the data will be represented. Including the option `eig = TRUE` additionally provides the eigenvalues of the data. They are used to calculate P and P' in order to decide on the dimensions of the projection space p. In this case, both criteria give satisfactory results, i.e. values greater than 0.8, for any number of dimensions. Given this result, it is convenient to set p equal 2 to plot the MDS map in a simple diagram.

```
> fit = cmdscale(d, eig = TRUE, k = 2)                    # fit mds model
```

If it is necessary to depict the results in a two-dimensional space, the following R commands can be used. The results are displayed in Fig. 8.6. It is evident that the car models form clusters. On the right side, one can see expensive luxury cars. On the left part of the plot, one finds affordable cars.

```
> plot(fit$points, type = "n")                    # set the plotting frame
> abline(v = 0, lty = "dotted")                   # y1 = 0 line
> abline(h = 0, lty = "dotted")                   # y2 = 0 line
> text(fit$points, labels = rownames(mydata))# add text
```

8.4.2 Non-metric Multidimensional Scaling

Unfortunately, classical metric MDS cannot always be used and other methods of scaling might be more suitable. This section is concerned with non-metric multidimensional scaling, which can be applied if there is a considerable number of negative eigenvalues and classical scaling of the proximity matrix may be inadvisable. Non-metric scaling is also used in the case of ordinal data, e.g. when comparing a range of colours. Customers might be able to specify that one was 'brighter' than another, without being able to attach any quantitative value to the extent the colours differ.

Non-metric MDS uses only a rank order of the proximities to produce a spatial representation of them. Thus, the solution is invariant under monotonic transformations of the proximities. One such method was originally suggested by Shepard (1962) and Kruskal (1964). Beginning from arbitrary coordinates in a p-dimensional space, e.g. calculated by metric MDS, the distances are used to estimate disparity between the objects using monotonic regression. The aim is to represent the fitted distances d_{ij}^y as $d_{ij}^y = \hat{d}_{ij}^x + \varepsilon_{ij}$, where the estimated disparities $\hat{d}_{ij}^x$ are monotonic with the observed proximities and, subject to this constraint, resemble the d_{ij}^y as closely as possible. For a given set of disparities, the required coordinates can be found by minimising some function of the squared differences between the observed proximities and the derived disparities, generally known as stress. The procedure is iterated until some convergence criterion is satisfied. The number of dimensions is chosen by comparing stress values or other criteria, for example R^2.

Non-metric MDS can be applied using R as well. The next example uses the voting data of the package HSAUR2. This dataset represents the voting results of 15 congressmen from New Jersey on 19 environmental bills.

To perform non-metric MDS, load the data, compute the distance matrix and run function isoMDS with the default two-dimensional solution.

```
> require(MASS)
> data(voting, package = "HSAUR2")        # load the data
> fit = isoMDS(voting)                     # fit MDS
> plot(fit$points, type = "n")             # plot the model
> abline(v = 0, lty = "dotted")            # y = 0 line
> abline(h = 0, lty = "dotted")            # x = 0 line
> text(fit$points, labels = rownames(voting))  # add text
```

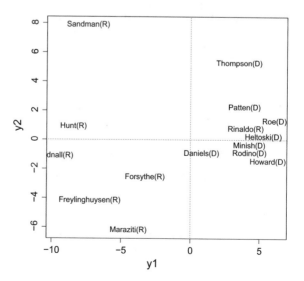

Fig. 8.7 Configuration of the MDS of voting data.
Q BCS_isoMDS

Figure 8.7 shows the output of the above procedure. It is clear that the Democratic congressmen have voted differently from the Republicans. A possible further conclusion is that the Republicans have not shown as much solidarity as their Democratic colleagues. More examples on MDS are given in Everitt and Hothorn (2011).

8.5 Discriminant Analysis

One of the most applied tools in multivariate data analysis is classification. Discriminant analysis is concerned with deriving rules for the allocation of observations to sets of a priori defined classes in some optimal way. It requires two samples—the training sample, for which group membership is known with certainty a priori, and the test sample, for which group membership is unknown.

The theory of discriminant analysis states that one needs to know the class posteriors $P(G \mid X)$, where G is a given class and X contains other characteristics of an object. Suppose $f_k(x)$ is the class-conditional density and let π_k be the prior probability of class k. A simple application of the Bayes' theorem gives

$$P(G = k \mid X = x) = \frac{f_k(x)\,\pi_k}{\sum_{j=1}^{K} f_j(x)\pi_j}.$$

It is easy to see that in terms of the ability to classify, the $f_k(x)$ is almost equivalent to having the quantity $P(G = k \mid X = x)$.

Linear discriminant analysis (LDA) arises in the special case when each class density is a multivariate Gaussian and classes have a common covariance matrix $\Sigma_k = \Sigma, \forall k$. The purpose of LDA is to find the linear combination of individual variables which gives the greatest separation between the groups. To discriminate between two classes k and l, a decision rule can be constructed as the log ratio

$$\log \frac{P(G = k \mid X = x)}{P(G = l \mid X = x)} = \log \frac{\pi_k}{\pi_l} - \frac{1}{2}(\mu_k - \mu_l)^\top \Sigma^{-1}(\mu_k - \mu_l) + x^\top \Sigma^{-1}(\mu_k - \mu_l),$$

(8.5)

where Σ, μ_k and μ_l are in most cases unknown and have to be estimated from the training data set. Equation (8.5) is linear in x.

The decision rule can also be expressed as a set of k linear discriminant functions

$$\delta_k(x) = x^\top \Sigma^{-1}\mu_k - \frac{1}{2}\mu_k^\top \Sigma^{-1}\mu_k + \log \pi_k.$$

An observation is assigned to the class with the highest value in the respective discriminant function. The parameter space $\mathbb{R}^p$ is divided by hyperplanes into regions that are classified as classes $1, 2, \ldots, K$.

In cases where the classes do not have a common covariance matrix, the decision boundaries between each pair of classes are described by a quadratic function. The corresponding quadratic discriminant functions are defined as

$$\delta_k(x) = -\frac{1}{2} \log |\Sigma_k| - \frac{1}{2}(x - \mu_k)^\top \Sigma_k^{-1}(x - \mu_k) + \log \pi_k.$$

For more details on discriminant analysis, see Hastie et al. (2009).

To illustrate the method and its implementation in R, the datasets `spanish` and `spanishMeta` are used. They contain information about the relative frequencies of the 120 most frequent tag trigrams (combination of three letters) in 15 texts contributed by three Spanish authors (Cela, Mendoza and Vargas Llosa). The aim of the analysis is to construct a classification rule which allows automatic assignment of a text by an 'unknown author' to Cela, Mendoza or Vargas Llosa. In this dataset the number of variables, i.e. the different tag trigrams, is much larger than the number of observations. In addition, some of the variables are highly correlated. Practically, this means that much information that is conveyed by the variables is redundant. We therefore can perform a principal component analysis before constructing the discriminant function, in order to reduce dimensions without much loss of information.

```
> require(MASS)
> data(spanish, package ="languageR")      # load the data
> mydata    = t(spanish)                    # transpose
> pca       = prcomp(mydata,                # fit PCA model
+     center = TRUE,                         # center values
+     scale  = TRUE)                         # and rescaled
> datalda = pca$x
> datalda = datalda[order(rownames(datalda)), ] # sort by rownames
> data(spanishMeta, package ="languageR")   # load data
> mydata = cbind(datalda[,1:2], spanishMeta$Author)
> colnames(mydata) = c("PC1","PC2","Author")
> mydata    = as.data.frame(mydata)
> mydata$Author = as.factor(mydata$Author)
```

Before performing LDA, the dataset is randomly divided into a training dataset and a test dataset. The training dataset is used to construct a discrimination rule, which is subsequently applied to the test dataset in order to test its precision. Performing the precision test on the same data for which the classification rule was constructed is bad practice, since the results are not reliable, e.g. biased towards overfitting.

Alternatively, a wide range of resampling techniques such as bootstrap and cross-validation can be used. These methods are described in Hastie et al. (2009).

```
> # set.seed(123)                           # set seed, see Chap.\,9
> n                = nrow(mydata); n        # total number of observations
[1] 15
> nt               = floor(0.6 * n); nt     # set training set size
[1] 9
> indices = sample(1:n, size = nt)          # sample
> mydata.train = mydata[indices, ]          # define the training set
> mydata.test  = mydata[-indices, ]         # define the test set
```

To perform LDA in R, one can use the `lda` function in the package MASS. The output are the prior probabilities for each group, the group means, the coefficients of the linear discriminant functions and the proportion of the trace, i.e. which proportion of variance is explained by each discriminant function.

```
> fit = lda (Author ~ PC1 + PC2,     # fit LDA model
+     data = mydata.train)
> fit
Call:
lda(Author ~ PC1 + PC2, data = mydata.train)

Prior probabilities of groups:
    1    2    3
0.33 0.22 0.44

Group means:
   PC1    PC2
1 -3.3  -4.50
2  4.3   4.27
3  1.7  -0.83

Coefficients of linear discriminants:
       LD1    LD2
PC1  -0.26  -0.14
PC2  -0.22   0.12

Proportion of trace:
   LD1    LD2
0.9929 0.0071
```

Having used this classification rule, one can easily depict discrimination borders for the groups to see how distinguishable the three classes are, see Fig. 8.8. The borders between two classes are obtained by the difference between the corresponding discriminant functions. For this purpose use function `partimat` from package `klaR`.

Fig. 8.8 LDA for the Spanish author data. Q BCS_LDA

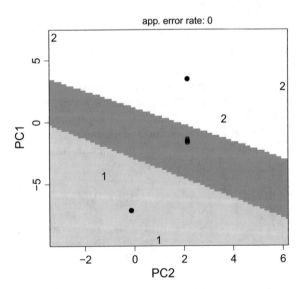

```
> require(klaR)
> partimat(Author ~ PC1 + PC2,   # multiple figure array
+    data   = mydata_test,        # for dataset
+    method = "lda",              # using LDA
+    main   = " ")                # no title
```

Predicted classes and posterior probabilities can be obtained by using the function `predict()`. The table below shows the probability of each element of falling into each of the classes.

```
> pred.class = predict(fit, mydata.test)$class    # predicted class
> pred.class
[1] 1 1 3 2 3 3
Levels: 1 2 3
> post.prop = predict(fit, mydata.test)$posterior # posterior prob.
> post.prop
              1      2      3
X14459g11  0.75   0.00   0.24
X14460g11  0.97   0.00   0.03
X14464g11  0.00   0.75   0.25
X14466g11  0.01   0.28   0.72
X14467g11  0.02   0.25   0.73
X14474g11  0.12   0.07   0.81
```

To check whether the discrimination rule works well, the percentage of correctly classified observations is calculated. In this case, the percentage of error is 33%, which can be interpreted as high or low depending on the application. The calculations of the prediction error are shown below.

```
> pr.table = table(mydata_test$Author, pred.class) # pred. vs true
> pr.table
   pred.class
    1 2 3
  1 2 0 0
  2 0 1 2
  3 0 0 1
> pred.correct = diag(prop.table(pr.table, 1))
> pred.correct
   1    2    3
1.00 0.33 1.00                                      # prediction in %

> 1 - sum(diag(prop.table(pr.table)))              # prediction error
[1] 0.3333333
```

Chapter 9
Random Numbers in R

*Anyone who considers arithmetical methods of producing
random digits is, of course, in a state of sin.*

— John von Neumann

Random number generation has many applications in economic, statistical, and financial problems. With the advantage of high speed and cheap computation, new statistical methods using random number generation have been developed. Important
examples are the bootstrap based procedures. When referring to a random number
generator of any statistical software package, the phrase 'random number' is misleading, as all random number generators are based on specific mathematical algorithms.
Thus, the computer generates deterministic and therefore pseudorandom numbers,
which are called 'random' for simplicity. In this context, the standard uniform distribution plays a key role, because its random numbers can be transformed so as
to obtain pseudo-samples from any other distribution. True random numbers can be
obtained by sampling and processing a source of natural entropy such as atmospheric
noise, radioactive decay, etc.

The main purpose of this chapter is to provide some computational algorithms
that generate random numbers.

9.1 Generating Random Numbers

A sequence of numbers generated by an algorithm is entirely determined by the
starting value of the algorithm, often called the *seed* or *key*. While the determinism
of the random numbers generated might be considered a drawback, it is also an
important property in simulation and modeling, due to the ability to repeat the process
using the same seed value. In addition, these algorithms are more efficient, as they can
produce many numbers in a short time. In simulation and especially in cryptography,

© Springer International Publishing AG 2017
W.K. Härdle et al., *Basic Elements of Computational Statistics*,
Statistics and Computing, DOI 10.1007/978-3-319-55336-8_9

huge amounts of random numbers are used, thus sampling speed is crucial. Typically, the algorithms are periodic, which means that the sequence repeats itself in the long run. While periodicity is hardly ever a desirable characteristic, modern algorithms have such long periods that they can be ignored for most practical purposes.

Definition 9.1 (*Pseudorandom Number Generator*) A pseudorandom number generator is a structure $\Xi = (S, s_0, T, U, G)$, where S is a finite set of states, s_0 is the initial state, also called the 'seed' or 'key', $T : S \to S$ is a transformation function, U is a finite set of output symbols, and $G : S \to U$ is the output function.

The initial state of the generator is s_0 and it evolves according to the recurrence $s_n = T(s_{n-1})$, for $n = 1, 2, 3, \ldots$. At step n, the generator creates $u_n = G(s_n)$ as output. For $n \geq 0$, the u_n are the random numbers produced by the generator. Due to the fact that S is finite, the sequence of states s_n is eventually periodic. So the generator must eventually reach a previously seen state, which means $s_i = s_j$ for some $0 \leq i < j$. This implies that $s_{j+n} = s_{i+n}$ and therefore $u_{j+n} = u_{i+n}$ for all $n \geq 0$. The length of the period is the smallest integer p such that $s_{p+n} = s_n$ for all $n \geq r$ for some integer $r \geq 0$. The smallest r with this property is called transient. For $r = 0$, the sequence is called purely periodic. Note that the length of the period cannot exceed the maximal number of possible states $|S|$. Thus a good generator has p very close to $|S|$. Otherwise, this would result in a waste of computer memory.

9.1.1 Pseudorandom Number Generators

Modular arithemtic is often used to cope with the issue of generating a sequence of apparently random numbers on computer systems, which are completely predictable. The basic relation of modular arithmetic is called *equivalence modulo m*, where m is an integer. The modulo operation finds the remainder of the division of one number by another, e.g. $7 \bmod 3 = 1$. As stated in Sect. 1.4.1, the modulo operator in R is %%.

In the following, we present two pseudorandom number generators, which illustrate the main ideas behind such algorithms.

Linear congruential generator

The *Linear Congruential Generator* (LCG) is one of the first developed and best-known pseudorandom number generator algorithms. It is fast and can be easily implemented.

Definition 9.2 (*Linear Congruential Generator*) The LCG is a recursive algorithm
of the form

$$T(x_i) = (ax_{i-1} + c) \bmod m, \text{ with } 0 \le x_i < m \text{ for } i = 0, 1, 2, \ldots,$$
$$\text{and with } m > 0, 0 < a < m, 0 \le c < m, 0 \le x_0 < m,$$

where m, a, c and x_0 are the modulus, multiplier, increment and seed value, respectively.

To obtain numbers with the desired properties discussed in Sect. 9.1, one has to
transform the generated integers into [0, 1] with

$$G(x_i) = \frac{x_i}{m} = U_i, \quad \text{for } i = 0, 1, 2, \ldots.$$

The selection of values for a, c, m and x_0 drastically affects the statistical properties and the cycle length of the generated sequence of integers. The full cycle length
is m if and only if

1. $c \ne 0$,
2. c and m are relatively prime, i.e. their greatest common divisor is 1,
3. $a - 1$ is divisible by all prime factors of m,
4. and if m is divisible by 4, $a - 1$ also has to be divisible by 4.

In addition, Marsaglia (1968) has shown that these points, when plotted in n-
dimensional space, will lie on at most $m^{1/n}$ hyperplanes, see Fig. 9.1. This is illustrated
by a famous example of badly chosen starting values, namely in RANDU, a random
number generator developed by IBM. This algorithm was first introduced in the early
1960s and became widespread soon after.

Definition 9.3 (*RANDU—The IBM Random Number Generator*) RANDU is a Linear Congruential Generator defined by the recursion

$$T(x_i) = (2^{16} + 3)x_{i-1} \bmod 2^{31}, \text{ with } 0 \le x_i \le m \text{ and } i = 0, 1, 2, \ldots.$$

The corresponding R code is

```
> RANDU = function(n, seed = 1){
+     x = NULL                      # predefine constants
+     a = 2^16 + 3
+     m = 2^31
+     for(i in 1:n){
+         seed = (a * seed) %% m
+         x[i] = seed / m           # normalise the values to [0, 1]
+     }
+     x
+ }
> RANDU(4)
[1] 3.051898e-05 1.831097e-04 8.239872e-04 3.295936e-03
```

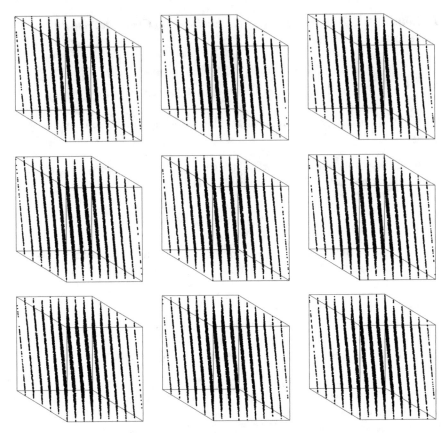

Fig. 9.1 Nine plots of random numbers x_{k+2} versus x_{k+1} versus x_k generated by RANDU visualised: in a three dimensional space all points fall in 15 hyperplanes. ⊘ BCS_RANDU

The chosen modulus, $m = 2^{31}$, is not a prime and the multiplier was chosen primarily because of the simplicity of its binary representation, not for the goodness of the resulting sequence of integers. Consequently, RANDU does not have full cycle length and has some clearly non-random characteristics.

To demonstrate the inferiority of these values, consider the following calculation where mod 2^{31} has been omitted from each term.

$$
\begin{aligned}
x_{k+2} &= (2^{16} + 3)x_{k+1} = (2^{16} + 3)^2 x_k \\
&= (2^{32} + 6 \cdot 2^{16} + 9)x_k = \{6 \cdot (2^{16} + 3) - 9\}x_k \\
&= 6x_{k+1} - 9x_k
\end{aligned}
$$

The linear dependency between x_{k+2}, x_{k+1} and x_k is obvious. According to Marsaglia's Theorem, all points fall in 15 hyperplanes in a three dimensional space, as illustrated in Fig. 9.1. Today, many results depending on computations with RANDU from the '70s are seen as suspect.

Lagged fibonacci generator

Another idea is to use the Fibonacci numbers on moduli like $x_i \bmod m$, where $x_i = x_{i-1} + x_{i-2}$ with $x_0 = 0$ and $x_1 = 1$. Unfortunately, this sequence does not have satisfactory randomness properties. One solution is to combine terms at greater distances, in other words, to add a lag between the summands. This is the main idea of the Lagged Fibonacci Generator (LFG).

Definition 9.4 (*The Lagged Fibonacci Generator*) A *Lagged Fibonacci Generator* is a recursive algorithm defined as

$$T(x_i) = (x_{i-j} + x_{i-k}) \bmod m,$$

with $0 \le x_i \le m$, $i, j, k = 0, 1, 2, \ldots$ and $0 < k < j < i$.

Unlike the LCG, the seed is not a single value. It is rather a sequence of (at least) j integers, of which one integer should be odd. The statistical properties of the resulting sequence of numbers rely heavily on this seed.

The value of the modulus m does not by itself limit the period of the generator, as it does in the case of a LCG. The maximum cycle length for $m = 2^M$ is $(2^j - 1) \cdot 2^{M-1}$ if and only if the trinomial $x^l + x^k + 1$ is primitive over the integers mod 2.

Using the notation $LFG(j, k, p)$ to indicate the lags and the power of two moduli, a commonly used version of this algorithm is $LFG(17, 5, 31)$. The cycle length of this version is 2^{47}.

```
> LFG = function(j, k, p, n) {
+    seed = runif(j, 0, 2^p)                         # generate the seed
+    for(i in 1:n) {
+       seed[j + i] = (seed[i] + seed[j + i - k]) %% 2^p
+    }
+    seed[(j + 1):length(seed)] / max(seed) # standardise to [0, 1]
+ }
> LFG(17, 5, 31, 4)                                  # generate 4 random numbers
[1] 0.3102951 0.9048108 0.4415016 1.0000000
```

The basic problem with this generator is, that there exist three-point correlation between x_{i-k}, x_{i-j} and x_i given by the construction of the generator itself, but typically these correlations are very small.

Mersenne twister

The Mersenne twister is a pseudorandom number generator developed by Matsumoto and Nishimura (1998). Due to its good properties, it is widely used even nowadays. Its name is derived from the fact that the period length is a Mersenne prime, i.e. a prime which is one less than a power of two: $M_p = 2^p - 1$

This section presents the most common version of this algorithm, also called $MT19937$.

Definition 9.5 (*The Mersenne Twister 'MT19937'*) The sequence of numbers generated by the MT 19937 is uniformly distributed on $[0, 1]$. To save computation time, the generator works internally with binary numbers. The main equation of the generator is given by

$$x_{k+n} = x_{k+m} + x_{k+1}A \begin{pmatrix} 0 & 0 \\ 0 & I_r \end{pmatrix} + x_k A \begin{pmatrix} I_{w-r} & 0 \\ 0 & 0 \end{pmatrix}, k = 0, 1, \ldots,$$

where x_i is a 32 dimensional row vector, I_r a 19×19 identity matrix, I_{w-r} a 13×13 identity matrix, $n = 351$, $m = 175$ and

$$A = \begin{bmatrix} 0 & 1 & 0 & \ldots & 0 \\ 0 & 0 & 1 & \ldots & 0 \\ \vdots & \vdots & \vdots & \ddots & \vdots \\ 0 & 0 & 0 & \ldots & 1 \\ a_{31} & a_{30} & a_{29} & \ldots & a_0 \end{bmatrix}.$$

As a result, each vector x_i is a binary number with 32 digits. Afterwards, the resulting vector x_{k+n} is rescaled. $x_0 = 4357$ is chosen to be the most appropriate seed.

To explain the recursion above, one can think of it as a concatenation and a shift: a new vector is generated by the first 13 entries of x_k and the last 19 entries of x_{k+1}. The shift results from the multiplication by A, which is in some way disturbed by the addition of $a_0, a_1, \ldots$. The result is added to x_{k+m}.

The resulting properties of the generated sequence of numbers are extremely good. The period length of $2^{19937} - 1$ ($\approx 4.3 \cdot 10^{6001}$) is astronomically high and sufficient for nearly every purpose today. It is k-distributed to 32-bit accuracy for every $1 \le k \le 623$ (see Sect. 9.3.2). In addition, it passes numerous tests for statistical randomness.

9.1.2   Uniformly Distributed Pseudorandom Numbers

To generate a sequence of uniformly distributed pseudorandom numbers on (min, max) in R, the command `runif()` is used, see Sect. 4.2.

```
> runif(5, 0, 1)    # runif(number of observations, min, max)
[1] 0.9388026 0.6177511 0.1474307 0.1756104 0.3917517
```

The underlying algorithm of `runif()` is the Mersenne twister, discussed in Sect. 9.1.1. `runif()` will not generate either of the extreme values, unless *max* = *min* or *max* − *min* is small compared to *min*, and in particular not for the default arguments:

```
> min(runif(100000))                    > max(runif(100000))
[1] 8.260133e-06                        [1] 0.9999601
```

As already mentioned, the sequences generated by `runif()` are the result of a pseudorandom generator, which therefore rely on a seed. R uses the predefined seed by default.

```
> .Random.seed[1:5]
[1] 403 83 -1212313168 -168900013 -1327450767
```

`.Random.seed` is an integer vector, containing the seed for random number generation. Due to the fact that all implemented generators use this seed, it is strongly recommended **not to alter this vector!**

One can define a specific starting value with the function `set.seed()`. This is a great way to ensure that simulation results are reproducible by using the same seed value, as shown in the following.

```
> set.seed(2)      # fix the seed
> x1 = runif(5)
[1] 0.1848823 0.7023740 0.5733263 0.1680519 0.9438393
> x2 = runif(5)
[1] 0.9434750 0.1291590 0.8334488 0.4680185 0.5499837
> set.seed(2)      # use the same seed value as for x1
> x3 = runif(5)
[1] 0.1848823 0.7023740 0.5733263 0.1680519 0.9438393
> x1 == x2         # comparison of the generated sequences
[1] FALSE FALSE FALSE FALSE FALSE
> x1 == x3
[1] TRUE TRUE TRUE TRUE TRUE
```

`set.seed()` uses its single integer argument to automatically set as many seeds as required for the pseudorandom number generator. This is considered a simple way of getting quite different seeds by specifying small integer arguments, and also a way of getting valid seed sets for the more complicated methods.

9.1.3 Uniformly Distributed True Random Numbers

In contrast to pseudorandom generators, the R package `random` provides users with a source of true randomness that comes from www.random.org. Since 1998, the site has been offering true random numbers generated from an atmospheric noise sample via a radio tuned to an unused broadcast frequency combined with a skew correction originally due to John von Neumann. This method might be better suited for some purposes than pseudorandom number generators, but its speed of obtaining random numbers is generally relatively low. Using the package or website and its database is therefore a little more time consuming.

```
> require(random)
> x = randomNumbers(n = 1000, min = 1, max = 100, col = 1) / 100
> head (as.vector(x))
    [,1] [,2] [,3] [,4] [,5] [,6]
V1 0.75 0.08 0.02 0.94 0.43 0.78
```

Obviously, the specification of `set.seed()` is of no use in this context, as the sequences of random numbers are always truly random and not reproducible.

Due to the slow generation of these numbers, one reasonable method is to use these random numbers to generate a *seed* for further algorithms.

9.2 Generating Random Variables

In contrast to random number generation, rv generation always refers to the generation of variables whose probability distribution is different from that of the uniform distribution on $(0, 1)$. The basic problem is therefore to generate an rv X whose distribution function is assumed to be known.

rvs invariably use a random number generator as their starting point, which yields a uniformly distributed variable on $(0, 1)$ (see Sect. 9.1.1). The technique is to manipulate or transform one or more such uniform rvs in an elegant and efficient way to obtain a variable with the desired distribution.

As with all numerical techniques, there is more than one method available to generate variables for the desired distribution. Four factors should be considered when selecting an appropriate generator:

1. *Exactness* refers to the distribution of the variables produced by the generator. A generator is said to be *exact* if the distribution of the variables generated has the exact form of the desired distribution. In some situations where the accurate distribution is not critical, methods producing an approximate distribution may be acceptable.
2. *Speed* refers to the computing time required to generate a variable. There are two contributions to the overall time: the *setup time* to create the constants or calculating tables, and the *variable generation time*. The importance of these two contributions depends on the application. If a sequence of rvs, all obeying the same distribution, is needed, then the time needed to set up the tables and constants only counts once, because the same values can be used for every variable of the sequence. If each variable has a different distribution, the setup time is just as important as the variable generation time.
3. *Space* refers to the computer memory requirements of the generator. Some algorithms make use of extensive tables which can become significantly costly if different tables need to be held in memory simultaneously.
4. *Simplicity* refers to the elementariness of the algorithm.

9.2.1 *General Principles for Random Variable Generation*

In this chapter, the three main principles for rv generation, the *inverse transform method*, *acceptance–rejection method*, and the *composition method*, will be discussed. For simplicity, it is assumed that there is available a pseudorandom number generator that produces a sequence of independent $U(0, 1)$ variables, as discussed in Sect. 9.1.1

The inverse transform method

From the property of the quantile function given in Definition 4.11, which states that for $U \sim U(0, 1)$, the rv $X = F^{-1}(U)$ has cdf F, i.e. $X \sim F$, one can create rvs very efficiently whenever F^{-1} can be calculated. This method is called the inverse transform method.

Recall, that the inverse transform method has been shown to work even in the case of discontinuities in $F(x)$. As a result, the generated X will satisfy $P(X \leq x) = F(x)$, so that X has the required distribution.

The acceptance–rejection method

Suppose the inverse of F is unknown or numerically hard to calculate and one wants to sample from a distribution with pdf $f(x)$. Under the following two assumptions, the acceptance–rejection method can be used:

1. There is another function $g(x)$ that dominates $f(x)$ in the sense that $g(x) \geq f(x) \; \forall x$.
2. It is possible to generate uniform values between 0 and $g(x)$. These values will be either above or below $f(x)$.

Definition 9.6 (*The Acceptance–Rejection Method*)

1. Generate $U_1 \sim U\left[supp\,(f)\right]$, where $f(x)$ is the pdf of F and $supp(f)$ is the support of f.
2. Generate $U_2 \sim U[0, g(U_1)]$.
3. If $U_2 < f(U_1)$, return U_1 (the x-coordinate) as the generated X value, otherwise repeat the procedure.

It is intuitively clear that X has the desired distribution because the density of X is proportional to the height of f (Fig. 9.2).

The dominating function $g(x)$ should be chosen in an efficient way, so that the area between $f(x)$ and $g(x)$ is small, to keep the proportion of rejected points small. Additionally, it should be easy to generate uniformly distributed points under $g(x)$.

Fig. 9.2 The
Acceptance–Rejection
method. ⊙ BCS_ARM

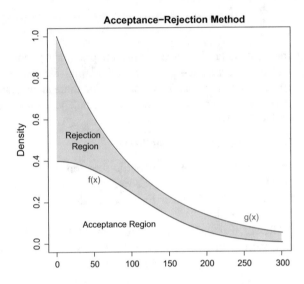

The average number of points (X, Y) needed to produce one accepted X is called the *trials ratio*, which is always greater than or equal to unity. The closer the trials ratio is to unity, the more efficient is the generator. To present a handy way of constructing a suitable $g(x)$, consider the density $f(x)$ of a distribution for which an easy way of generating variables already exists, and define $g(x) = K \cdot h(x)$. It can be shown that if X is a variable from $g(x)$ and U is uniformly distributed on $(0, 1)$ and independent of X, then the points $(X, Y) = \{X, K \cdot U \cdot h(x)\}$ are uniformly distributed under the graph of $g(x)$. In this case, K has to be chosen in such a way that $g(x) \geq f(x)$ is assured. Therefore the trials ratio is exactly K.

The composition method

Suppose a given density f can be written as a weighted sum of n densities

$$f(x) = \sum_{i=1}^{n} p_i \cdot f_i(x),$$

where the weights p_i satisfy the two conditions $p_i > 0$ and $\sum_{i=1}^{n} p_i = 1$. In such a framework, the density f is said to be a *compound* density. This method can be used to split the range of X into different intervals, so that sampling from each interval facilitates the overall process.

9.2.2 *Random Variables*

For several distributions, R provides predefined functions for generating rvs. Most of these functions will be discussed later in this chapter, to give a general overview of this field. The syntax in this area follows a straight structure. All commands are compositions of d, p, q, r (which stand for the density, distribution function, quantile function, and rvs), plus the name of the desired distribution, as discussed in Chaps. 4 and 6. Thus

```
> rbinom()
```

will give the pdf of the binomial distribution, and

```
> rexp()
```

will give an rv from the exponential distribution with parameters set by default.

9.2.3 *Random Variable Generation for Continuous Distributions*

This section discusses the generation of rvs for several continuous distributions. Starting with three famous algorithms for the normal distribution, several ways of generating rvs for the exponential, gamma, and beta distribution will be presented.

The normal distribution

As already mentioned briefly in Sect. 4.3 rnorm() produces n rvs for the normal distribution with mean equal to 0 and standard deviation equal to 1 by default.

```
> rnorm(n, mean = 0, sd = 1)

> # generate 4 observations from a N(0, 1)
> rnorm(4)
[1] -0.3936441 -0.1939292  0.1383921  0.4417582
> # change the default algorithm
> RNGkind(normal.kind = "Box-Muller")
> # generate 4 observations from a N(0, 1)
> rnorm(4)
[1] -0.1367969  1.3994082 -0.3936441 -0.1939292
```

A famous method developed by Box and Muller (1958) was the earliest method for generating normal rvs and, thanks to its simplicity, it was used for a long time. The algorithm is provided in the following definition.

Definition 9.7 (*The Box–Muller Method*) Let U_1 and U_2 be independent rvs obeying the uniform distribution $U(0, 1)$. Consider the rvs

$$X_1 = \sqrt{-2 \log(U_1)} \cos(2\pi U_2)$$
$$X_2 = \sqrt{-2 \log(U_1)} \sin(2\pi U_2)$$

Then X_1 and X_2 are independent and standard normally distributed, i.e. $(X_1, X_2)^\top \sim N(0, \mathcal{I}_2)$. Considering $\Sigma^{1/2}(X_1, X_2)^\top$, we have dependent rvs, see Fig. 6.4.

Unfortunately, this algorithm is rather slow due to the fact that for each number a square root, log, and a trigonometric function have to be computed.

Neave (1973) has shown that the Box–Muller method shows a large discrepancy between observed and expected frequencies in the tails of the normal distribution when U_1 and U_2 are generated with a congruential generator. This effect became known as the *Neave effect* and is a result of the dependence of the pairs generated by a congruential generator, such as RANDU. This problem can be avoided by using two different sources for U_1 and U_2, as shown in the following R code.

```
> boxmuller = function(n){
+    if(n %% 2 == 0){a = n / 2}else{a = n / 2 + 1}
+    x1 = x2 = 1:a
+    for (i in 1:a) {
+       u1    = runif(1)                              # generate two
+       u2    = runif(1)                              # uniform rvs
+       x1[i] = sqrt(-2 * log(u1)) * cos(2 * pi * u2) # transformation
+       x2[i] = sqrt(-2 * log(u1)) * sin(2 * pi * u2)
+    }
+    c(x1, x2)                                        # print results
+ }
> boxmuller(4)
 [1] 2.527755 -1.548469 -0.794818 -1.777311
```

Marsaglia (1964) mentioned a slightly different version of the Box–Muller method, in which the trigonometric function was replaced to reduce the computation time. It is known as the polar method and can be seen as an accelerated version of the algorithm in Definition 9.7.

Definition 9.8 (*The Polar Method*) Generate two independent observations u_1, u_2 from a uniform distribution on $(-1, 1)$ and set $w = u_1^2 + u_2^2$. If $w > 1$, repeat these steps, otherwise set $z = \sqrt{(-2 \log w)/w}$ and define $x_1 = u_1 z$ and $x_2 = u_2 z$. Then $(x_1, x_2)^\top$ should be an observation from $N(0, \mathcal{I}_2)$.

```
> polarmethod = function(n){
+    if(n %% 2 == 0){a = n / 2}else{a = n / 2 + 1}
+    x1 = x2 = 1:a                   # create output variables X and Y
+    i  = 1                          # set counter
+    while(i <= a) {
+       u1 = runif(1, -1, 1)         # generate two uniform random numbers
+       u2 = runif(1, -1, 1)
+       w  = u1^2 + u2^2
+       if (w <= 1) {
+          z     = sqrt((-2 * log(w)) / w)
+          x1[i] = u1 * z
```

```
+        x2[i]  = u2 * z
+         i     = i + 1              # precede counter
+        }
+     }
+     c(x1, x2)                      # print results
+  }
> polarmethod(8)
[1]  0.41867423   0.90550395  -0.07986714   1.17828848
[5]  0.65455600  -0.71171498  -0.05401868   0.90767865
```

The first part produces a point (u_1, u_2) which is an observation from an rv uniformly distributed on $[0, 1]^2$. If w is smaller than 1, this point is located inside the unit circle. Then $u_1/\sqrt{w}$ is equivalent to the sine and $u_2/\sqrt{w}$ to the cosine of a random direction (angle). Moreover, the angle is independent of w, which is an observation of the rv that follows uniform distribution. This method is a good example of the acceptance–rejection method for the normal distribution.

Definition 9.9 (*Ratio of Uniforms*) Generate u_1 from $U(0, b)$, u_2 from $U(c, d)$ and $x = u_1/u_2$, with $b = \sup \{h(x)\}^{1/2}$, $c = -\sup x \{h(x)\}^{1/2}$ and $d = \sup x \{h(x)\}^{1/2}$.

$$\text{If} \begin{cases} u_1^2 \le h(u_2/u_1), & \text{deliver } x; \\ \text{otherwise,} & \text{repeat the algorithm;} \end{cases}$$

where $h(\cdot)$ is some density function.

For the normal distribution with the non-normalised density $h(x) = \exp(-x^2/2)$, the algorithm can be stated as follows.

Generate u_1 from $U(0, 1)$ and u_2 from $U(-\sqrt{2/e}, \sqrt{2/e})$, where e is the base of the natural logarithm. Let $x = u_2/u_1$ and $z = x^2$.

$$\text{If} \begin{cases} z \le 5 - \{4 \exp(1/4)\} u_1, & \text{deliver } x \text{ (Quick accept);} \\ z > \{4 \exp(-1/4)\}/u_1 - 3, & \text{repeat the algorithm (Quick reject);} \\ z \le -4 \log u_1, & \text{deliver } x; \\ \text{otherwise,} & \text{repeat the algorithm.} \end{cases}$$

Given these conditions, we can define an acceptance region

$$
\begin{aligned}
C_h &= \left\{ (u_1, u_2) : 0 \le u_1 \le h^{1/2}(u_2/u_1) \right\} \\
&= \left\{ (u_1, u_2) : 0 \le u_1 \le \sqrt{\exp\{-u_2^2/(2u_1^2)\}} \right\} \\
&= \left\{ (u_1, u_2) : (u_2/u_1)^2 \le -4 \log u_1 \right\}.
\end{aligned}
$$

The inequality can then be stated in terms of the variable x. To avoid repeated computation of $\log u_1$, the inner and outer bounds defined by the following inequalities on $\log u$ are calculated.

$$(4 + 4 \log c) - 4cu_1 \le -4 \log u_1,$$
$$-4 \log u_1 \le 4/cu_1 - (4 - 4 \log c)$$

These two inequalities arise from the fact that the tangent line, taken at the point d, lies above the concave log function:

$$\log y \le y/d + (\log d - 1).$$

Taking $y = u_1$ and $d = 1/c$ leads to the lower bound, using $y = 1/u_1$ and $d = c$ yields the upper bound. Note that the area of the inner bound is largest when $c = \exp(1/4)$ and note that the constant $4 \cdot \exp(1/4) = 5.1361$ is computed and stored in advance to avoid computing the same trigonometric function a second time.

The exponential distribution

The first algorithm provided in this section is interesting, because it uses only arithmetic operations.

Definition 9.10 (*Neumann's Algorithm*) Generate a number of random observations $u_1, u_2, \ldots$ from a uniform distribution as long as their values consecutively decrease, i.e. until $u_{n+1} > u_n$. If n is even, return $x = u_n$, otherwise repeat the procedure.

```
> neumannmeth = function(n){
+    x     = 1:n            # erase used variables
+    dummy = 0
+    i     = 1
+    while (i <= n){
+       us = runif(1)           # generate two uniform random numbers
+       ug = runif(1)
+       l  = 2                  # set the even-counter
+       while (us < ug) {
+          dummy = us           # save smaller random number
+          us    = ug           # overwrite smaller number
+          ug    = runif(1)     # generate new uniform
+          l     = l + 1        # precede counter
+       }
+       # if l is even, save dummy in x and precede counter
+       if ((l %% 2 == 0) && (dummy != 0)){
+          x[i]  = dummy
+          i     = i + 1
+          dummy = 0
+       }
+    }
+    x
+ }
> neumannmeth(10)
 [1] 0.59064976 0.62533392 0.61146766 0.92082615 0.66926057
 [6] 0.06609662 0.09082758 0.76548550 0.55991772 0.60223848
```

Despite its simplicity, this method should not be used for generating rvs, because too many uniformly distributed random numbers are needed to generate one exponentially distributed rv. Neumann's rather inefficient algorithm can be improved by applying the result of Pyke (1965)'s Theorem. It states that $n\left(U_{(k+1)} - U_{(k)}\right) =$

$nS_k \sim \mathcal{E}(k)$, where $U_{(1)} \le U_{(2)} \le \ldots \le U_{(n)}$ is an ordered series of standard uniformly distributed rvs with $S_k = U_{(k+1)} - U_{(k)}$ and $S_n = U_{(n)}$.

One important advantage in computing rvs from the exponential distribution is the fact that the exponential distribution has a closed form expression for the inverse of its cumulative distribution function. Given a random number generator, some numbers X must be selected, which need to obey an exponential distribution. The following definition states the selection procedure.

Definition 9.11 (*Inverse cdf Method for the Exponential Distribution*) First, generate a variable u from $U(0, 1)$, then calculate $x = 1 - \exp(-\lambda \cdot u)$.

```
> invexp = function(n, lambda){
+   sapply(1:n, function(x) {1 - (exp(-lambda * runif(1)))})
+ }
> invexp(9, 2)
 [1] 0.8499208 0.4120199 0.5673652 0.4386219 0.4934604 0.6861557
 [7] 0.3059538 0.5698566 0.7037824
```

`rexp()` uses the algorithm by Ahrens and Dieter (1972), which is faster than the inverse method presented above.

```
> ptm = proc.time()
> invexp(10000, 1)
> proc.time() - ptm
       User    System elapsed
       0.39      0.00    0.66

> ptm = proc.time()
> rexp(10000)
> proc.time() - ptm
       User    System elapsed
       0.04      0.00    0.04
```

The gamma distribution

The command

```
> rgamma(n, shape, rate = 1, scale = 1 / rate)
```

generates n rvs with shape parameter b, default rate of 1, and scale parameter $1/rate$. For $b \ge 1$, a specific algorithm by Ahrens and Dieter (1982b) is used, but for $0 < b < 1$, the following, different, algorithm by Ahrens and Dieter (1974) is used.

Definition 9.12 (*The Acceptance–rejection Method of* Ahrens and Dieter 1982b)

1. Generate u_1 from $U(0, 1)$ and set $w = u_1 \cdot (e + b)/e$, where e is the base of the natural logarithm.
2. If $w < 1$, go to 3, else to 4.

3. Generate u_2 from $U(0, 1)$ and set $y = w^{1/b}$. If $u_2 \leq \exp(-y)$, return $x = ay$, else go to 1.
4. Generate u_2 from $U(0, 1)$ and set $y = -\log\left\{(\frac{e+b}{e} - w)/b\right\}$. If $u_2 \leq y^{1/b}$, return $x = ay$, else go to 1.

Definition 9.13 (*The Acceptance–rejection Method of* Cheng 1977)

1. Generate u_1 and u_2 from $U(0, 1)$.
2. Set $v = (2b - 1)^{-1/2} \log(u_1/1 - u_2)$, $y = b\exp(v)$, $z = u_1^2 u_2$,
   and $w = b - \log(4) + b + (2b - 1)^{1/2}v - y$.
3. If $w + 1 + \log(4.5) - 4.5z \geq 0$ or $w \geq \log(z)$ holds, then return $x = ay$, else go to 1.

Note that the *trials ratio* improves from $4/e \approx 1.47$ to $\sqrt{(4/\pi)} \approx 1.13$ as $b \to \infty$. The *setup time* is rather short, as only four constants have to be computed in advance.

Definition 9.14 (*The Acceptance–rejection Method of* Fishman 1976)

1. Generate u_1 and u_2 from $U(0, 1)$.
2. Set $v_1 = -\log(u_1)$ and $v_2 = -\log(u_2)$.
3. If $v_2 > (b - 1)\{v_1 - \log(v_1) - 1\}$, return $x = av_1$, else go to 1.

This algorithm was introduced by Atkinson and Pearce (1976) and is simple and short. It is also efficient for $b < 5$, as the trials ratio is reduced from unity at $b = 1$ and to 2.38 at $b = 5$. Note that for greater values of b, Algorithm 9.13 by Cheng is more efficient.

The beta distribution

The function

```
> rbeta(n, shape1, shape2, ncp = 0)
```

generates n rvs of the beta distribution with the two shape parameters p and q and a default non-centrality parameter of 0. `rbeta()` is based on the following algorithm by Cheng (1978).

Definition 9.15 (*The Acceptance-rejection Method of* Cheng 1978)

1. Generate u_1 and u_2 from $U(0, 1)$.
2. Set $v = \sqrt{(p + q - 2)/(2pq - p - q)} \log\{u_1/(1 - u_1)\}$ and $w = p\exp(v)$.
3. If $p + q\log(p + q)/(q + w) + p + \sqrt{(p + q - 2)/(2pq - p - q)}^{-1}v - \log(4)$
   $< \log(u_1^2 u_2)$, go to 1.
4. Else return $x = w/(q + w)$.

This method has a bounded trials ratio of less than $4/e \approx 1.47$.

9.2.4 *Random Variable Generation for Discrete Distributions*

The general methods of Sect. 9.2.3 are in principle available for constructing discrete variable generators. However, the special characteristics of discrete variables imply certain modifications.

The binomial distribution

In R, rvs from the binomial distribution, see 3.6, can be generated via

```
> rbinom(n, size, prob)
```

where *n* is the number of observations, *size* the number of trials, and *prob* the probability of success.

Definition 9.16

1. Set $x = 0$.
2. Generate *u* from $U(0, 1)$.
3. If $u \leq p$, set $y = 1$, else $y = 0$.
4. Set $x = x + y$.
5. Repeat *n* times from step 2, then return *x*.

This algorithm uses the fact that *x* is an observation from binomially distributed rv with *n* and *p*, i.e. the sum of *n* independent Bernoulli variables with parameter *p*. Note that the generation time increases early with *n*.

As pointed out earlier, the fastest binomial generators for fixed parameters *n* and *p* are obtained via table methods. On the downside for these methods, the memory requirements and the *setup time* for new values of *n* and *p* are proportional to *n*, which is a major drawback. More useful is a simple inversion without a table resulting in a short algorithm and a shorter setup time. The execution time is proportional to $n \cdot \min(p, 1 - p)$. Therefore, rejection algorithms were proposed because they are on the whole both fast and well suited for changing the values of *n* and *p*, as typically required in simulation.

The implemented algorithm for `rbinom()` is based on a version by Kachitvichyanukul and Schmeiser (1988). The algorithm generates binomial variables via an acceptance/rejection based on the function

$$f(x) = \frac{\lfloor np + p \rfloor! \cdot (n - \lfloor np + p \rfloor)!}{\lfloor x + 0.5 \rfloor!(n - \lfloor x + 0.5 \rfloor)!} \left(\frac{p}{1-p} \right)^{\lfloor x+0.5 \rfloor - \lfloor np+p \rfloor} \quad \text{for } -0.5 \leq x \leq n + 0.5.$$

The resulting algorithm dominates other algorithms with constant memory requirements when $n \cdot \min(p, 1 - p) \geq 10$ in terms of execution times. Only for $n \cdot \min(p, 1 - p) \leq 10$ is the inverse transformation algorithm faster. An implementation of the inverse transformation is presented below.

```
> bininv = function(num, n, p){
+     x = 1:n
+     for(i in 1:n){
+         q = 1 - p                    # setup constants
+         s = p / q
+         a = (n + 1) * s
+         r = q ^ n
+         y = 0
+         u = runif(1)                 # generate uniform variable
+         while(u > r){                # check condition
+             u = u - r
+             y = y + 1
+             r = ((a / y) - s) * r
+         }
+         x[i] = y
+     }
+     x
+ }
> bininv(5, 10, 0.5)
 [1] 3 5 7 6 5
```

The poisson distribution

Rvs from the Poisson distribution can be generated via

```
> rpois(n, lambda)
```

where n is the number of observations and lambda the vector of (non-negative) means. The implemented algorithm was first mentioned by Ahrens and Dieter (1982a). The following algorithm was developed by Knuth (1969) and is a simple way of generating random Poisson distributed variables by counting the number of events that occur in a time period t.

Definition 9.17 *(Knuth's Algorithm)*

1. Set $p = 1$, $k = 0$ and let u be an observation from the unifor distribution.
2. If $\exp(-\lambda) < p$, set $k = k + 1$ and $p = p \cdot u$, else print out $k - 1$.

```
> rpoisson = function(lambda = 1){
+     L = exp(-lambda)
+     k = 0
+     p = 1
+     while(p > L){
+         k = k + 1
+         p = p * runif(1, 0, 1)
+     }
+     k - 1
+ }
```

The advantages of this algorithm are that only one constant $L = \exp(-\lambda)$ has to be evaluated and that it requires only a minimum amount of storage space. However, the time to generate an rv increases rapidly with λ. The following method can be used for large λ, such as $\lambda > 25$, based on the fact that the distribution of $\lambda^{-1/2}(X - \lambda) \xrightarrow{\mathcal{L}}$ $N(0, 1)$. Bear in mind that this is an asymptotic result.

```
> poisson.as = function(n, lambda = 1){
+     a = lambda^0.5
+     sapply(1:n,
+         function(x){max(0, trunc(0.5 + lambda + a * rnorm(1)))})
+ }
> poisson.as(10, 30)
 [1] 31 27 31 35 33 34 34 26 27 31
```

A comparison of the computation times for both algorithms with $\lambda = 45$ and $n = 10000$ is shown below.

```
> proc.time(for (i in 1:10000) {pois[i] = rpoisson(45)})
       User       System elapsed
      95.39        13.07  3931.30

> proc.time(poisson.as(10000, 45))
       User       System elapsed
      91.74        13.07  3927.62
```

9.2.5 Random Variable Generation for Multivariate Distributions

Variable generation is generally much more complicated for multivariate distributions than for univariate ones, excepting those multivariate distributions with independent components. The added complications arise from the dependencies between the components of the random vector, which must be dealt with in multivariate distributions. One general approach to creating such a dependency structure is the conditional sampling method.

Conditional sampling

The beauty of the conditional sampling approach is that it reduces the problem of generating a p-dimensional random vector into a series of p univariate generation tasks.

Definition 9.18 (*Conditional Sampling*) Let $X = (X_1, X_2, \ldots, X_d)^\top$ be a random vector with joint distribution function $F(x_1, x_2, \ldots, x_d)$. Suppose the conditional distribution of X_j, given that $X_i = x_i$, for $i = 1, 2, \ldots, j - 1$, is known for each j.

Then the vector X can be built up one component at a time, where each component is obtained by sampling from a univariate distribution and recusrsively calculating each $X_1, X_2, \ldots, X_d$.

For this method, it is necessary to know all the conditional densities. Therefore, its usefulness depends heavily on the availability of the conditional distributions and, of course, on the difficulty of sampling from them.

The transformation method

If the conditional distributions of X are difficult to derive, then perhaps a more convenient transformation can be found. The key element for this method is to represent X as a function of other, usually independent, univariate rvs. An example of this method is the Box–Muller method (see Definition 9.7), which uses two independent uniform variables and converts them into two independent normal variables.

Even though the transformation method has wide applicability, it is not always trivial to find a transformation with which to generate a multivariate distribution of a given X. The following guidelines by Johnson (1987) have proven helpful.

1. Beginning with the functional form $f_X(x)$, one could apply invertible transformations to the components of X in order to find a recognizable distribution.
2. Consider transformations of X that simplify arguments of transcendental functions in the density $f_X(x)$.

The rejection method for multivariate distributions

Obviously, the rejection method presented in Sect. 9.2.1 is not tied to a particular dimension. But even though the theory carries over straightforwardly from the univariate to the multivariate case, there exist some significant practical difficulties. As stated by Johnson (1987), the main problem is to find a dominating function $g_X(x)$ for $f_X(x)$ if the dependence among the components of X is strong. Intuitively, one could choose a density for $g_X(x)$ which corresponds to the independent components with the same marginal distributions as X. But in most cases, as the dependencies in X increase, the extent to which $g_X(x)$ approximates $f_X(x)$ decreases. Therefore the *trials ratio* approaches infinity. In general, the design of an efficient rejection method is more difficult than in the univariate case.

If the standard rejection method for multivariate distributions is inapplicable due to the above computational difficulties, a random vector from high-dimensional distributions can be generated by Markov Chain Monte Carlo (MCMC) techniques. The MCMC algorithms, like the Metropolis–Hastings algorithm or the Gibbs sampler, have therefore become standard tools for Bayesian econometricians. Note, as a critical remark, that the generated sample is a Markov chain and that even random vectors of small samples are not necessarily iid. For a detailed review, we refer to Albert (2009) and to Martin et al. (2011) on using the `MCMCpack` package.

The composition method for multivariate distributions

Like the rejection method, the composition method is not tied to a specific space like $\mathbb{R}^1$. A method for obtaining dependence from independence is the following.

Definition 9.19 Define a random vector $X = (X_1, \ldots, X_d)^\top$ as $(SY_1, \ldots, SY_d)$, where the Y_i are iid rvs and S is a random scale. In such a framework, the distribution of X is a scale mixture. The resulting density $f_X(x)$ of X is

$$f_X(x) = E\left[\prod_{i=1}^{d}\left\{\frac{f_Y}{S}\left(\frac{x_i}{S}\right)\right\}\right].$$

Simulating from copula-based distributions

There are numerous methods of simulating from copula-based distributions, see Frees and Valdez (1998), Whelan (2004), Marshall and Olkin (1988), McNeil (2008), Scherer and Mai (2012). The conditional inverse method is a general approach aimed at simulating rvs from an arbitrary multivariate distribution. Here we sketch this method with an example of simulating from copulae. We use conditional sampling to generate rvs $U = (u_1, \ldots, u_d)^\top$ recursively from the a sample $v_1, \ldots, v_d$ from a uniformly distributed rvs $V_1, \ldots, V_d \sim U(0, 1)$ and the conditional distributions. We set $u_1 = v_1$. The rest of the variables are calculated using the recursion $u_i = C_i^{-1}(v_i|u_1, \ldots, u_{i-1})$ for $i = 2, \ldots, d$, where $C_i = C(u_1, \ldots, u_i, 1, \ldots, 1)$ and the conditional distribution of U_i is given by

$$C_i(u_i|u_1, \ldots, u_{i-1}) = P(U_i \le u_i|U_1 = u_1 \ldots U_{i-1} = u_{i-1})$$
$$= \frac{\partial^{i-1}C_i(u_1, \ldots, u_i)}{\partial u_1 \ldots \partial u_{i-1}} \Big/ \frac{\partial^{i-1}C_{i-1}(u_1, \ldots, u_{i-1})}{\partial u_1 \ldots \partial u_{i-1}}.$$

The method is numerically expensive, since it depends on higher order derivatives of C and the inverse of the conditional distribution function.

Simulating from archimedean copulae

The idea of the Marshal–Olkin method is based on the fact that the Archimedean copulae are derived from Laplace transforms. Let M be the univariate cdf of a positive rv (so that $M(0) = 0$) and let ϕ be the Laplace transform of M, i.e.,

$$\phi(s) = \int_0^\infty \exp\{-sw\}\, dM(w), \text{ with } s \ge 0.$$

For any univariate distribution function F, a unique distribution G exists, given by

$$F(x) = \int_0^\infty G^\alpha(x)\, dM(\alpha) = \phi\{-\log G(x)\}.$$

Considering d different univariate distributions $F_1, \ldots, F_d$, we obtain

$$C(u_1, \ldots, u_d) = \int_0^\infty \prod_{i=1}^d G_i^\alpha \, dM(\alpha) = \phi\left[\sum_{i=1}^d \phi^{-1}\{F_i(u_i)\}\right],$$

which is a multivariate distribution function.

One proceeds with the following three steps to make a draw from a distribution described by an Archimedean copula:

1. Generate an observation u from M;
2. Generate observations $(v_1, \ldots, v_d)$ from R;
3. The generated vector is computed by $x_j = G_j^{-1}(v_j^{1/u})$.

This method works faster than the conditional inverse technique. The drawback is that the distribution M can be determined explicitly only for a few generator functions ϕ, for example the Frank, Gumbel and Clayton families.

A simple implementation in R makes use of the package `copula`, which was discussed in detail in Sect. 6.3. The command `rMvdc()` draws n random numbers from a specified copula.

```
> # specification of the Clayton copula with uniform marginals
> require(copula)
> uniclayMVD = mvdc(claytonCopula(0.79),
+     margins     = c("unif", "unif"),
+     paramMargins = list(list(min = 0, max = 1),
+         list(min = 0, max = 1)))
> # 10000 random number draw from the Clayton copula
> rMvdc(uniclayMVD, n = 10000)
```

Figure 9.3 shows 10,000 random numbers drawn from a Clayton copula.

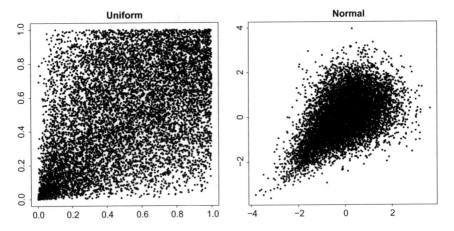

Fig. 9.3 10,000 realizations of an rv with uniform marginals in $[0, 1]$ (*left*) and with standard normal marginals (*right*) with the dependence structure in both cases given by a Clayton copula with $\theta = 0.79$. **Q** BCS_claytonMC

9.3 Tests for Randomness

The first tests for random numbers in history were published by Kendall and Smith (1938). They were built on statistical tools, such as the Pearson chi-square test, which were developed in order to distinguish whether or not experimental phenomena matched up with their theoretical probabilities.

Kendall and Smith's (SJP) original four tests were hypothesis tests, which tested the null hypothesis that each number in a given random sequence had an equal chance of occurring, and that various other patterns in the data should also be distributed equiprobably. The four tests are:

- The *frequency test* is a very basic test which checks whether there are roughly the same number of 0's, 1's, 2's, 3's, etc.
- The *serial test* does the same for sequences of two digits at a time (00, 01, 02, etc.), comparing their observed frequencies with their hypothetical predictions based on equal distribution.
- The *poker test* is used to test for certain sequences of five numbers at a time (00000, 00001, 00011, etc.), based on hands in the game poker.
- The *gap test* looks at the distances between zeroes (00 would be a distance of 0, 030 would be a distance of 1, 02250 would be a distance of 3, etc.).

In general, it is very hard to check if some sequence is truly random—just consider a single number, say 7. The reason is that if the random number generator is good, each and every possible sequence of values is equally likely to appear, as Kendall and Smith stated in 1938.

To illustrate this fact, consider a coin toss experiment with 3 throws. If the coin is fair, the resulting sequences of heads and tails are all equally likely, i.e., $P(H, T, T) = P(H, H, T) = P(T, T, T)$, and so on. Therefore, even a fair coin can generate a sequence of only heads or tails. And even worse: that sequence appears with the same probability as any other sequence, which may appear much random! This means that a good random number generator will also produce sequences that look non-random to the human eye and which fail any statistical tests that we might apply to it. Therefore, it is impossible to prove that a given sequence of numbers is random.

How to proceed, if it is impossible to conclusively prove randomness? We can follow a pragmatic approach by taking many sequences of random numbers from a given generator and testing each one of them. One can expect that some of the sequences will fail some of the tests. As the sequences pass more of the tests, the confidence in the randomness of the numbers increases and with it the confidence in the generator. However, if many sequences fail the tests, we should be suspicious.

This is also the way one would intuitively test a coin to see if it is fair: throw it many times, and if too many sequences of the same value come up, one should be suspicious. The problem with randomness still tends to be the same: you can never be sure.

Nevertheless, in the following sections, two of the less intuitive and hard to pass tests will be provided, in contrast to the more natural approaches above.

9.3.1 Birthday Spacings

The birthday spacing test is one of a series of tests called Diehard tests, which were developed by Marsaglia (1995) and published on a CD. Consider the following situation. If m birthdays are randomly chosen from a year of n days (usually 365) and sorted, the number of duplicate values among the spacings between those ordered birthdays will be asymptotically Poisson distributed with parameter $\lambda = m^{\frac{3}{4n}}$.

Theory provides little guidance on the speed of the approach to the limiting form, but extensive simulation with a variety of random number generators provides values of m and n for which the limiting Poisson distribution seems satisfactory. Among these are $m = 1024$ birthdays for a year of length $n = 2^{24}$ with $\lambda = 16$.

9.3.2 k-Distribution Test

A more strenuous theoretical requirement for a generator is to be "k-distributed". A sequence is 1-distributed if every number it generates occurs equally often, 2-distributed if every pair of numbers occurs equally often, and so on. In a fair coin-flip context, 1-distribution would mean that heads and tails occurred equally often, and 2-distribution would mean that all results of two tosses occurred equally often.

Definition 9.20 A sequence x_i of integers of period P is said to be k-distributed to v-bit accuracy if k of the kv-bit vectors is

$$\{\text{trunc}_v(x_i), \ \text{trunc}_v(x_{i+1}), \ ..., \ \text{trunc}_v(x_{i+k-1})\} \quad \text{for } 0 \leq i < P,$$

where $\text{trunc}_v(x)$ denotes the number formed by the leading v bits of x, i.e. trunc_2 $(0.23917) = 0.23$, and each of the 2^{kv} possible combinations of bits occurs the same number of times in a period, except for the all-zero combination that occurs less often by one instance.

To test for k-distribution to n-bit accuracy, at least 2^{kn} measurements are needed. Thus, this property is generally shown theoretically without preforming the actual measurements. Nevertheless, it is possible to test for small k. In the case of $k = 2$, each pair of the sequence $\{U_i, U_{i+1}\}_{i=1}^{2n-1}$ refers to certain points of the unit square, where $U_i \sim U(0, 1)$. Decomposing the unit square into n^2 subsquares and counting the number of points in the subsquares allows using a χ^2-test for independence, since the number of observed and expected points in each cell can be compared. This example can be extended to larger k.

For large k and n, an alternative is to count the number of missing k-outcomes in a long string produced by the generator. The resulting count should be approximately normally distributed with a certain mean and variance, which must be determined by theory and simulation.

Finally, the most common random number generators generally cannot claim any better than a 1-distribution.

Chapter 10
Advanced Graphical Techniques in R

> *All children are artists. The problem is how to remain an artist once he grows up.*
>
> —Pablo Picasso

Data visualisation is an important part of data analysis with R. The standard Renvironment has various graphical facilities for drawing different types of statistical plots. However, there exist several shortcuts not covered in the base Rgraphical system. For example, basic Rprovides no capabilities for interactive plotting, including rotation and zoom of the existing plot. Moreover, dynamic graphics, e.g. on-fly adding of information and adjustment of parameters in the plot, are not embedded either.

For this reason, these important components are discussed within the add-on packages `rgl` and `rpanel`. Apart from this, the `lattice` package considerably extends the functionality of R by implementing multipanel plots, which are very useful for more precise multivariate data analysis.

In this chapter, we discuss these three important add-on packages for advanced data visualisation and provide relevant examples.

10.1 Package `lattice`

The `lattice` add-on package is an implementation of `Trellis` graphics in R. `Trellis` graphics, originally implemented in S and S-Plus at AT&T Bell Laboratories, is a data visualisation framework developed by Becker, Cleveland and Shyu. It provides powerful visualisation tools, see Becker et al. (1996). Here we only discuss the most important features of the `lattice` package, whereas considerably more detailed description is offered by the developer of the lattice system, Sarkar (2010).

© Springer International Publishing AG 2017
W.K. Härdle et al., *Basic Elements of Computational Statistics*,
Statistics and Computing, DOI 10.1007/978-3-319-55336-8_10

The name `Trellis` comes from the trellis-like rectangular array of panels similar to a garden trellis. By means of the `Trellis` graphics it is possible to study the dependence of a response variable on more than two explanatory variables. *Multipanel conditioning* is used for displaying multiple plots in one page with shared coordinate scales, aspect ratios and labels. This feature is especially useful for plotting multivariate and panel data, and is not provided by the standard Rgraphic system.

The design goal of the `Trellis` system is the optimisation of the available output area, therefore `Trellis` graphics provide default settings that produce superior plots in comparison to its traditional counterparts.

The `lattice` package is based on the `grid` graphics system, which is a low-level graphics system, see Sarkar (2010). `grid` does not provide high-level functions to create complete plots, but creates a basis for developing high-level functions as well as facilitates the manipulation of graphical output in `lattice`. Since `lattice` consists of `grid` calls, it is possible to add grid output to lattice output and vice versa, see R Development Core Team (2012). The knowledge of the `grid` package would be beneficial for customising the plots in `lattice`. Nevertheless `lattice` is a self-contained graphics system, enabling one to produce complete plots, functions for controlling the appearance of the plots and functions for opening and closing devices.

The short description of the package functions and relevant examples of the `lattice` graphical output will be given in the following.

10.1.1   Getting Started with `lattice`

The `lattice` package contains functions, objects and datasets. Most of the functions implemented in `lattice` are already available in the traditional Rgraphics environment. The complete list is given in Table 10.1.

Each of the listed high-level functions creates a particular type of display by default. Although the functions produce different output, they share many common features, i.e. that several common arguments affect the resulting displays in similar ways. These arguments are extensively documented in the help pages for `xyplot()`. The most important of them are the `formula` argument, describing the variables, and the `panel` argument, specifying the plotting function. These will be explained in more details in the following subsections.

10.1.2   `formula` Argument

The `Trellis` formula argument has a central role in `lattice`, since it is deployed in order to define statistical models. Different high-level generic functions employ several different types of notations. The most often used notations are presented in Table 10.2.

Table 10.1 High-level functions in `lattice`

`lattice` functions	Default display	Traditional functions
`barchart()`	Barplot	`barplot()`
`histogram()`	Histogram	`hist()`
`densityplot()`	Conditional kernel density plot	–
`dotplot()`	Dotplot	`dotchart()`
`bwplot()`	Comparative Box-and-Whisker plot	`boxplot()`
`stripplot()`	Stripchart	–
`qqmath()`	Theoretical quantile plot	`qqplot()`
`qq()`	Two-sample quantile plot	`qqplot()`
`dotplot()`	Cleveland dot plot	`dotchart()`
`xyplot()`	Scatter plot	`plot()`
`contourplot()`	Contour plot of surfaces	`contour()`
`cloud()`	3D Scatter plot	–
`levelplot()`	Level plot of surfaces	`image()`
`parallel()`	Parallel coordinates plot	`parcoord()`
`splom()`	Scatter plot matrix	`pairs()`
`wireframe()`	3D Perspective plot of surfaces	`persp()`

Table 10.2 Trellis formula notations

Notation	Explanation	Example of the function
$\sim x$	Plots a single variable x	`bwplot()`, `histogram()`, `qqmath()`
$y \sim x$	Plots variable y against variable x	`xyplot()`, `qq()`
$y \sim x * z$	Plots three variables	`levelplot()`, `cloud()`, `wireframe()`
$y \sim x \mid z$	Plots y against x for each level of z	`xyplot()`

In order to avoid mistakes in the use of the `formula` argument, it should be kept in mind that the syntax of the `formula` in `lattice` differs from that of `formula` used in the `lm()` linear model function, see Chap. 8.

The variable on the left side of " $\sim$ " is a dependent variable, while the independent variable(s) is (are) placed on the right side. For graphs of a single variable, only one independent variable needs to be specified in the first row of Table 10.2.

In order to define multiple dependent or independent variables, the sign '+' is placed between them. In case of multiple dependent variables, the formula would be assigned as $y_1 + y_2 \sim x$, so that the variables y_1 and y_2 are plotted against the variable x. In fact, $y_1 \sim x$ and $y_2 \sim x$ will be superposed in each panel. In a similar way, one can set multiple independent or both independent and dependent variables simultaneously as is implied in the code of Fig. 10.1 later in this chapter.

To produce conditional plots, the conditioning variable should be also specified in the `formula` argument, standing after the '|' symbol. When multiple conditioning

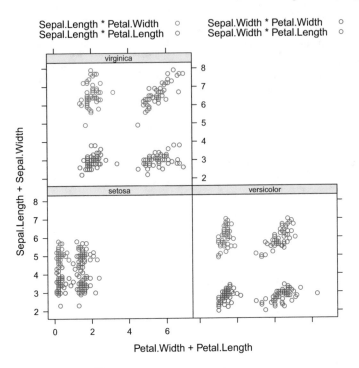

Fig. 10.1 Conditional plots. ✑ BCS_ConditionalGrouped

z variables are specified, then for each level combination of z_1 and z_2, lattice produces several plots of y against x, as depicted in Fig. 10.5. The notation is $y \sim x \mid z_1 + z_2$.

The definition of the formula argument is the initial step in the multilevel development process of lattice graphical output. The values used in the formula are contained in the argument data, specifying the data frame.

10.1.3 *panel Argument and Appearance Settings*

As mentioned above, the default settings of lattice plots are optimised for the traditional Rplots. The panel function is a function that uses a subset of the arguments to create a display. All lattice plotting functions have a default panel function, the name of which is built up from the prefix panel and the name of the function. For instance, the default panel function for the bwplot() function is panel.bwplot(). However, apart from superior default settings, lattice offers lots of flexibility due to its highly customisable panel functions.

There are two perspectives from which the lattice graph should be observed. First, the function call, e.g. histogram(), sets up the external components of the display, such as scale rectangle, axis lables, etc. Second, the panel function creates everything placed into the plotting region of the graph, such as plotting symbols.

The panel function is called from the general display function by the panel argument. Therefore, for the default settings, both function calls are identical.

```
> histogram(~x, data = dataset)
> histogram(~x, data = dataset, panel = panel.histogram)
```

There are different arguments that could be treated under the panel function. In order to temporarily change the default settings of, for instance, the plotting symbols, one can rewrite the new value into the panel function inside the general function call.

```
> xyplot(y ~ x,
+    data  = dataset,
+    panel = function(x, y){panel.xyplot(x, y, pch = 20)})
```

Alternatively, when it is desired to change the panel function arguments permanently, one should define a panel function as a separate function outside the general call function and then apply it to any function call.

```
> my.panel = function(x, y){panel.xyplot(x, y, pch = 29)}
```

Now by choosing the my.panel function, one would always use the type of the plotting points pch = 29.

In a similar way, different attributes (e.g. cex, font, lty, lwd, etc.) could be altered for a specific function either temporary or permanently.

10.1.4 Conditional and Grouped Plots

lattice offers conditional and grouped plots to work with and display multivariate data. In order to obtain a conditional plot, at least one variable should be defined as conditioning.

One gets different visual representations of the dataset, depending on whether the same variable is being used as conditioning or as grouping.

From the dataset iris, the conditioning variable Species is set to be conditioning, as shown in the Rcode below, which corresponds to Fig. 10.1.

```
> xyplot(Sepal.Length + Sepal.Width ~
+    Petal.Length + Petal.Width | Species,
+    data = iris)
```

Figure 10.1 contains three panels, standing for three types of Species. Each panel contains four combinations of iris characteristics.

Another alternative for displaying multivariate data is the groups argument. This splits the data according to the grouping variable. For the sake of comparability, Fig. 10.2 shows four panels, each one illustrating the combination of two variables and types of Species denoted by different colours.

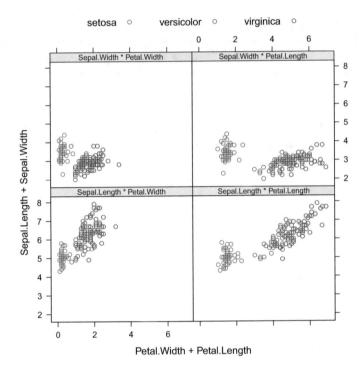

Fig. 10.2 Grouped plots. ⚆ BCS_ConditionalGrouped

```
> xyplot(Sepal.Length + Sepal.Width ~ Petal.Length + Petal.Width,
+    data   = iris,
+    groups = Species)
```

The use of a conditioning or grouping variable requires including a key legend in the graph. The argument `auto.key()` draws the legend, and the attribute `columns` defines the number of columns into which the legend is split.

According to this particular example, it is not very important how one employs the `Species` variable, since both outputs are qualitatively equal.

There are datasets where it is preferable to produce grouped plots rather than conditional plots. The following example of a density plot from the dataset `chickwts` confirms this.

```
> densityplot(~weight | feed,    # set conditional variable
+    data        = chickwts,
+    plot.points = FALSE)         # mask points
```

The resulting output is shown in Fig. 10.3. Since we employed the conditioning variable `feed`, which has six categories, Fig. 10.3 produces six panels with density plots.

Alternatively, the variable `feed` can be used as a grouping variable. Figure 10.4 creates one single panel with six superposed kernel density lines and enables a direct comparison between the different groups.

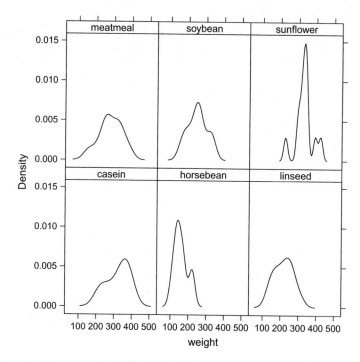

Fig. 10.3 Conditional density plots. 🔍 BCS_ConditionalGroupedDensity

```
> densityplot(~weight,
+    data        = chickwts,
+    groups      = feed,                    # set grouping variable
+    plot.points = FALSE,                   # mask points
+    auto.key    = list(columns = 3))       # legend in three columns
```

Moreover, the black and white colour scheme was applied to both plots on Fig. 10.3 as well as Fig. 10.4; in `lattice`, the colours will be changed by different types of symbols when the following code is applied.

```
> lattice.options(default.theme =
+ # set the default lattice color scheme to black/white scheme
+    modifyList(standard.theme(color = FALSE),
+ # set strips background to transparent
+       list(strip.background = list(col ="transparent")))))
```

10.1.5 Concept of `shingle`

`lattice` enables the use of continuous (numeric) variables as conditioning, with the `shingle` concept. A `shingle` is a data structure displaying the continuous

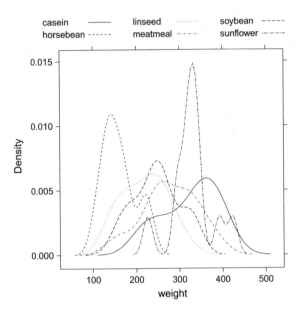

Fig. 10.4 Grouped and overlayed density plots. 🔍 BCS_ConditionalGroupedDensity

variables in the form of factors. It consists of a numeric vector and possibly overlapping intervals.

To convert a continuous variable into a `shingle` object means to split it into (possibly overlapping) intervals (levels). In order to do this, one uses the `shingle()` function, whereas the function `equal.count()` is used when splitting into equal length intervals is required. The `number` argument defines the number of intervals, whereas the `overlap` argument assigns the fraction of points to be shared by the consecutive intervals. The endpoints of the intervals are chosen in such a way that the counts of points in the intervals are as equal as possible. `shingle` returns the list of intervals of the numeric variable.

In the following Rcode, the continuous variables `temperature` and `wind` are split into four equal non-overlapping intervals and can be treated as usual factor variables. The new factor variables `Temperature` and `Wind` are considered as the conditioning variables.

```
> Temperature = equal.count(environmental$temperature,
+     number   = 3,                    # split into 3 equal intervals
+     overlap  = 0)                    # no overlapping
> Wind        = equal.count(environmental$wind,
+     number   = 4,
+     overlap  = 0)
> xplot(ozone ~ radiation | Temperature * Wind,
+     data      = environmental,
+     as.table  = TRUE)                # panels layout top to bottom
```

Figure 10.5 depicts the simultaneous use of these two conditioning variables. `Temperature` now contains three levels and `Wind` has four levels, though a rec-

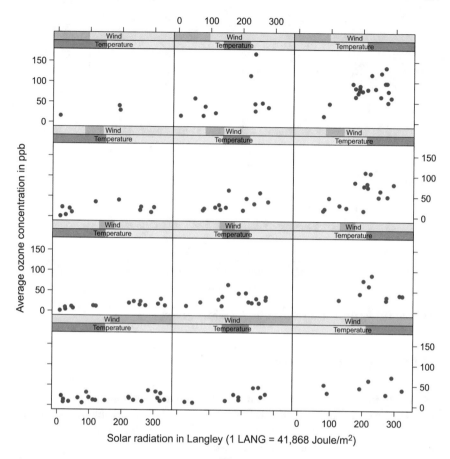

Fig. 10.5 Plot with two conditioning variables. **Q** BCS_TwoConditioningVariables

tangular array of 12 panels was created, depicting the ozone variable against the radiation variable for each combination of conditioning variables.

In the code, the argument par.strip.text() controls the text on each strip with the main components cex, col, font, etc.. By default, lattice displays the panels from bottom to top and left to right. By defining the argument as.table = TRUE the panels will be displayed from top to bottom.

Of course more than two conditioning variables are also possible, but the increasing level of complexity of the graphical output should be kept in mind.

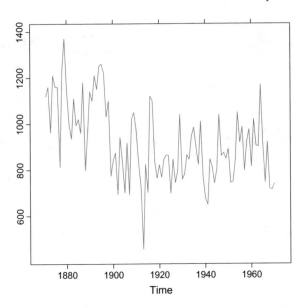

Fig. 10.6 Default time series plot. ☒ BCS_DefaultStackTimeSeries

10.1.6   Time Series Plots

The ability to draw multiple panels in one plot is particularly useful for time series
data. lattice enables cut-and-stack time series plots. The argument cut is specified by the number of intervals into which the time series dataset should be split, so
that changes over a time period can be studied more precisely.

The code for the simple time series plot in Fig. 10.6 is

```
> xyplot(Nile)
```

One can customise the plot by varying the arguments aspect, cut and strip,
where the last is responsible for the colour scheme of the strips.

```
> xyplot(Nile, aspect = "xy",
+     cut = list(number = 3,              # split into three panels
+         overlap = 0.1),                 # 10 per cent overlap
+     strip = strip.custom(bg = "yellow", # strips background
+         fg = "lightblue"))              # strips foreground
```

Figure 10.7 plots three panels, according to the number of intervals. Such a combination of two plots is most valuable for the user.

An object of class ts could also be a multivariate series, so that multiple time
series can be displayed in parallel in the same graph. For instance, by setting the
superpose argument to be TRUE, all series will be overlaid in one panel. When
the screens argument is specified, the series will be plotted into a predefined panel.

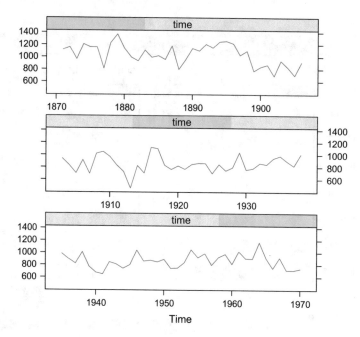

Fig. 10.7 Cut-and-stack time series plot. Q BCS_DefaultStackTimeSeries

10.1.7 Three- and Four-Dimensional Plots

The underlying philosophy of the Trellis system is to avoid three-dimensional displays and to use conditioning plots instead. Three-dimensional plots, created by wireframe() can be extended by a fourth variable used as conditioning. However, data interpretation appears to be more complicated, mostly because the plots are not rotatable.

For this reason, an analogous 3D plot will be constructed instead, with the rgl package (see Sect. 10.2). One can still create some superior three-dimensional plots with the levelplot() function. This function also allows upgrading a three-dimensional plot with another conditioning variable. The function levelplot() demands the dependent variable to be numerical and the conditioning variable, to be either factor or shingle. The following code shows how to create the 4D plot.

```
> levelplot(yield ~ site * variety | year,
+    data           = barley,
+    scales         = list(alternating = TRUE),
+    shrink         = c(0.3, 1),              # scale rectangles
+    region         = TRUE,
+    cuts           = 20,                     # range of dep. variable
+    col.regions    = topo.colors (100),  # color gradient
+    par.settings   = list(axis.text = list(cex = 0.5)),
+    par.strip.text = list(cex = 0.7),       # strips font size
+    between        = list(x = 1),           # space between panels
+    aspect ="iso", colorkey = list(space ="top"))
```

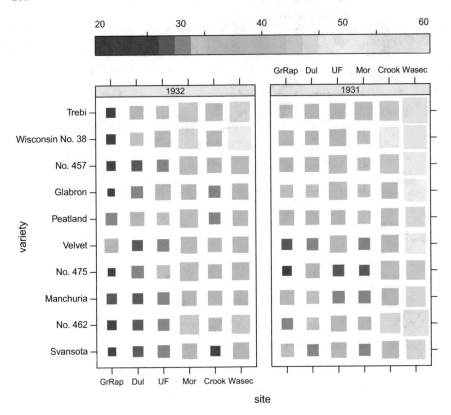

Fig. 10.8 Four-dimensional plot. ⊙ BCS_FourDimensional

The result of the listing is shown in Fig. 10.8, which presents the four-dimensional plot of the `lattice` data set `barley`. The explanatory variable `yield` is illustrated by means of the sizes and colours of the boxes. The higher the value of `yield`, the larger and lighter are the rectangles.

The arguments of interest are `cuts`, which specify the number of levels (the colour gradient) into which the range of a dependent variable is to be divided and `region`, which is a logical variable that defines whether the regions between the contour lines should be filled. Since `region = TRUE`, `col.regions` defines the colour gradient of the dependent variable. The `shrink` argument scales the rectangles proportionally to the dependent variable, and the `between` argument specifies the space between the panels on x and/or y axis.

The settings for the layout and appearance of the `lattice` plots facilitate an enhanced comprehension of the data. Multipanel conditioning is a central feature delivered by `lattice`, which enables data visualisation on multiple panels simultaneously, displaying different subsets of the data. Although `lattice` provides this kind of extended functionality, an immediate interactive control of the graphical output is still missing. For this reason, in the following chapter, we discuss the `rgl` and `rpanel` packages with an implemented interactive element.

10.2 Package `rgl`

RGL is a library of functions that offers three-dimensional real-time visualisation functionality with interactive viewpoint navigation in the Rprogramming environment. The RGL library is written in C++ using OpenGL. Since 3D objects need to be projected onto a 2D display, special navigation capabilities such as real-time rotation and zooming in/out are used to create the illusion of three-dimensionality. The further implementation in the RGL library of different features, including lighting, alpha blending, texture mapping, and fog effects, enhance this illusion, see Adler et al. (2003).

The `rgl` package includes both low-level `rgl.*` functions and a higher level interface for 3D rendering and computational geometry `r3d` with `*3d` functions. Most of the function calls exist also in a double set: both concepts from `rgl.*` and from `*3d` interfaces. The two principal differences between these are as follows: `rgl.*` calls set unspecified material properties to default values and `*3d` calls use the current values as defaults; `rgl.*` permanently changes the material properties with each call and `*3d` make temporary changes for the duration of the call.

The aim of this section is to give an overview of the `rgl` package structure, as well as some practical examples of the 3D real-time visualisation engine of the RGL library.

10.2.1 Getting Started with `rgl`

The `rgl` package can be subdivided into the following categories:

1. *Device management functions* include six functions, which control the RGL window device. These functions are used to open/close the device, to return the number of the active devices, to activate the device and to shut down the `rgl` device system or to re-initialise `rgl`.
2. *Scene management functions* enable stepwise removal of certain objects, such as shapes, lights, bounding boxes and background, from the 3D scene.
3. *Export functions* are used to save snapshots or screenshots in order to export them to other file formats.
4. *Environment functions* are set to alter the environment properties of the scene, e.g. to modify the viewpoint, background, axis labelling, bounding box, or to add a light source to the 3D scene.

5. The *appearance function* rgl.material(...) is responsible for the appearance properties, e.g. colour, transparency, texture and the sizes of the different object types.

We next demonstrate the shape functions in the rgl package combined with certain environment and object properties.

10.2.2 Shape Functions

The shape functions are an important part of the RGL library since they enable both the plotting of primitive shapes, such as points, lines, linestrips, triangles and quads, as well as high-level shapes, such as spheres and different surfaces, see Figs. 10.9, 10.10, 10.11, 10.12, 10.13 and 10.14.

RGL adds further shapes to the already opened device by default. To avoid this, one can create a new device window with the calls rgl.open() or open3d().

Fig. 10.9 Points.
Q BCS_Shapes

Fig. 10.10 Lines.
Q BCS_Shapes

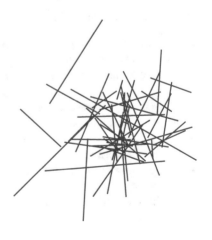

Fig. 10.11 Linestrips.
Q BCS_Shapes

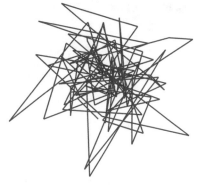

Fig. 10.12 Triangles.
Q BCS_Shapes

Fig. 10.13 Quads.
Q BCS_Shapes

Fig. 10.14 Spheres.
⌀ BCS_Shapes

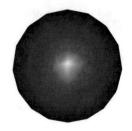

The shape functions in `rgl` are briefly described in the list below.

1. *3D points* are drawn by the function `rgl.points(x, y, z,...)`, see Fig. 10.9.
2. *3D lines* can be depicted with the function `rgl.lines(x, y, z,...)`, see Fig. 10.10. The nodes of the line are defined by the vectors x, y, z, each of length two.
3. *3D linestrips* are constructed with the function `rgl.linestrips(x,y, z,...)`. The nodes of the linestrips are, as in `rgl.lines(x, y, z,...)`, defined by the vectors x, y, z, each of length two. In the output, each next line strip starts at the point where the previous one ends, see Fig. 10.11.
4. *3D triangles* are created with the function `rgl.triangles(x, y, z,...)`, see Fig. 10.12. The vectors x, y and z, each of length three, specify the coordinates of the triangle.
5. *3D quads* can be drawn with the function `rgl.quads(x, y, z,...)`, see Fig. 10.13. The vectors x, y and z, each of length four, specify the coordinates of the quad.
6. *3D spheres* are not primitive, but they can be easily created with the function `rgl.spheres(x, y, z, r,...)`. This function plots spheres with centres defined by x, y, z and radius r. In order to create multiple spheres, one can define x, y, z, r as vectors of length n, see Fig. 10.14.
7. *3D surfaces* can be drawn by means of the generic `rgl.surface(x,...)` function. This is defined by a matrix specifying the height of the nodes and two vectors defining the coordinates.

Each of the shape functions can be produced with higher level functions from the r3d interface.

Alternatively, 3D surfaces can be constructed with the `persp3d(x,...)`, `surface3d()` or `terrain3d()` functions. As an example of a 3D surface, the hyperbolic paraboloid

$$\frac{x^2}{a^2} - \frac{z^2}{b^2} = y,$$

is produced by `surface3d()` and displayed in Fig. 10.15.

Fig. 10.15 Surface shape.
 BCS_SurfaceShape

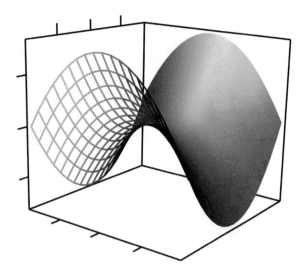

```
> require(rgl)
> x = z = -9:9
> f = function(x, z){(x^2 - z^2) / 10}
> y = outer(x, z, f)                          # square matrix,
>                                             # x rows, z columns
> open3d()
> surface3d(x, z, y,                          # plot 3D surface
+    back   ="lines",                         # back side grid
+    col    = rainbow(1000),
+    alpha = 0.9)                             # transparency level
> bbox3d(back ="lines", front ="lines")       # 3D bounding box
```

In this example of a 3D surface (see Fig. 10.15), a side dependent rendering effect
was implemented. This option gives the possibility of drawing the 'front' and 'back'
sides of an object differently. By default, the solid mode is applied, which can be
changed either to lines, points or cull (hidden). In Fig. 10.15, the front side is drawn
with a solid colour, whereas the back side appears to be a grid. This creates a better
illusion of 3D space. The bounding box is added to the scene with the function
bbox3d().

There are many options that can be used in order to make the 3D object look
more realistic. The lighting condition of the shape is described by the totality of light
objects. There are three types of lighting, i.e. the *specular* component determines the
light on the top of an object, *ambient* determines the lighting type of the surrounding
area and the *diffuse* component specifies the colour component, which scatters the
light in all directions equally. The light parameters specify the intensity of the light,
whereas theta and phi are polar coordinates defining the position of the light.

The following examples of 3D spheres (Figs. 10.16, 10.17, 10.18 and 10.19) depict
the effects of ambient and specular material on a sphere. Furthermore, the argument
smooth creates the effect of internal smoothing and determines the type of shading
applied to the spheres. When smooth is TRUE, Gouraud shading is used, otherwise

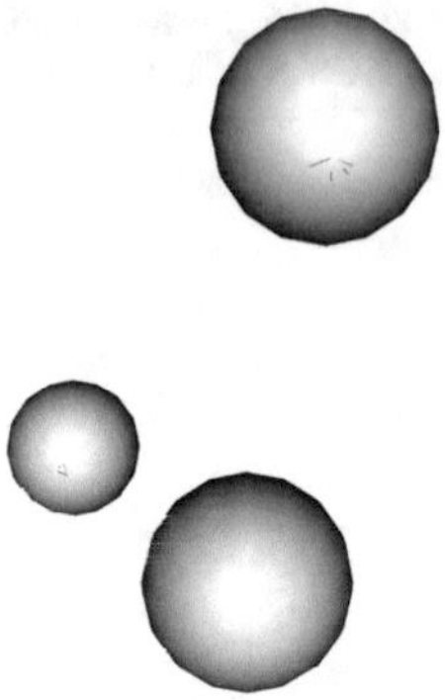

Fig. 10.16 Default light
source.
Q BCS_LightedPlots

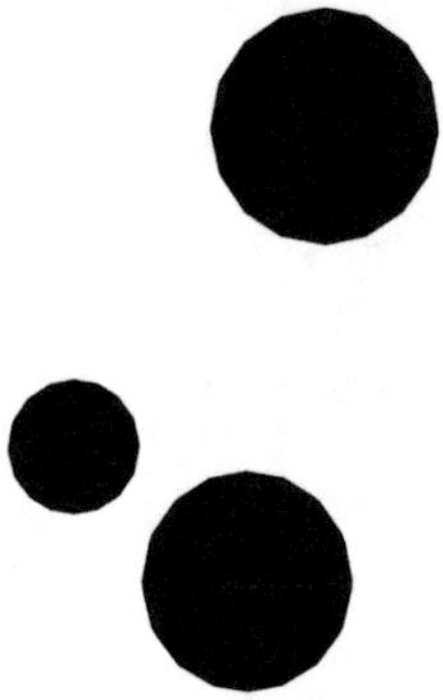

Fig. 10.17 Without light.
Q BCS_LightedPlots

Fig. 10.18 One light source
added. Q BCS_LightedPlots

flat shading. In Fig. 10.17, the `rgl.clear()` function is used in order to customise
the lighting scene of the display by deleting the lighting from the scene.

Fig. 10.19 Two light
sources added.
Q BCS_LightedPlots

```
> rgl.spheres(rnorm(3), rnorm(3), rnorm(3),   # set coordinates
+    r        = runif(5),                      # set radius
+    smooth = TRUE)                            # Gouraud shading
> rgl.clear(type ="lights")                    # remove the lighting
> # add the 1st light source
> rgl.light(theta = -90, phi = 50,             # position of light
+    ambient   ="white",                       # surrounding area
+                                              # lighting
+    diffuse   ="#dddddd",                     # diffusing lighting
+    specular ="white")                        # lighting on the top
> # add the 2nd light source
> rgl.light(theta = 45, phi = 30, ambient ="#dddddd",
+    diffuse ="#dddddd", specular ="white")
```

10.2.3 Export and Animation Functions

Exporting results from the `rgl` package differs from exporting classical graphical
outputs. For this reason, we will explain some of the main commands in this section.
To save the screenshot to a file in `PostScript` or in other vector graphics formats,
the function `rgl.postscript()` is used. There are also other supported formats,
such as `eps`, `tex`, `pdf`, `svg`, `pgf`. The `drawText` argument is a logical,
defining whether to draw text or not.

```
rgl.postscript("filename.eps", fmt ="eps", drawText = FALSE)
```

Alternatively, it is also possible to export the `rgl` content into bitmap `png` format
with the function `rgl.snapshot()`.

```
rgl.snapshot("filename.png", fmt ="png", top = TRUE)
```

The animation functions of the `rgl` package, such as `play3d()` or `movie3d()`,
are useful for demonstration purposes. `movie3d()` additionally records each single
frame to a `png` file. A movie in `gif` format can be produced by putting the created
`png` files into one document. Let us consider the example of a 3D surface. First,
define a 4×4 matrix `M` describing user actions to display the scene. Second, use

play3d(), where par3dinterp() returns a function which interpolates par3d parameter values, suitable for use in animations.

```
> M = par3d("userMatrix")              # 4x4 user actions matrix
> play3d(par3dinterp(userMatrix = list(M,
+    angle    = pi,                     # rotation angle
+    x = 1, y = 1, z = 0)),             # rotate around x and y axes
+    duration = 5)                      # duration of the rotation
```

By applying this code to the 3D surface example, one obtains a five second demonstration of the plot, rotated around the x- and y-axes.

Another alternative for manipulating the plot, rather than rotation and zoom, is provided by the function select3d(). This enables the user to select three-dimensional regions in a scene. This function can be used to pick out one part of the data, not influencing the whole dataset.

```
> if(interactive()){                       # interactive navigation
+                                           # is allowed
+    x = rnorm(5)                           # generate pseudo-random
+                                           # normal vector
+
+    y = z = x
+    r = runif(5)
+    open3d()                               # open new device
+    spheres3d(x, y, z, r, col = "red3")    # red spheres
+    k = select3d()                         # select the rectangle
+                                           # area
+    keep = k(x, y, z)                      # keep selected area
+                                           # unchanged
+    rgl.pop()                              # clear shapes
+    spheres3d(x[keep], y[keep], z[keep], r[keep],
+    col = "blue3")                         # redraw the selected
+                                           # area in blue
+    spheres3d(x[!keep], y[!keep], z[!keep], r[!keep],
+    col = "red3")                          # redraw the non-
+                                           # selected area in red
+
}
```

Fig. 10.20 Select a part of the scene. ⌕ BCS_SceneSelection

rgl.pop() is used for the same purpose as rgl.clear(), namely to remove the last added node on the scene. Figure 10.20 illustrates the selected spheres in blue, whereas the left data remains red.

To select the area, one draws a rectangle that represents the projection of the region onto the display. If the colour of the unselected area is not specified, its colour will be set to default after selection. A function which tests whether points are located in the selected region will be returned.

10.3 Package rpanel

The rpanel package employs different graphical user interface (GUI) controls to enable an immediate communication with the graphical output and provide dynamic graphics. Such an animation of graphs is possible by using single function calls, such as sliders, buttons or others, to control the parameters. If the particular state of a control button is altered, the response function call will be executed and the associated graphical display will be changed correspondingly.

rpanel is built on the tcltk package created by Dalgaard (2001). The tcltk package contains various options for interactive control offered by the Tcl/Tk system, whereas rpanel includes only a limited number of useful tools that enable the creation of control widgets through single function calls. rpanel offers the possibility of redrawing the entire plot, and interactively changing the values of the parameters set by the relevant controls, something which is not possible with the object-oriented graphics created, for instance, in Java, see Bowman et al. (2007). In order to be able to use the rpanel package, one should load the tcltk package first.

rpanel displays graphs in a standard Rgraphics window and creates a separate panel with control parameters. To avoid the necessity of operating with multiple panels, one can use the tkrplot package of Tierney (2005) to integrate the plot into the control panel, see Bowman et al. (2007).

10.3.1 Getting Started with rpanel

The rpanel package consists of control functions and application functions. The control functions are mainly used to build simple GUI controls for the Rfunctions. The most useful GUI controls are listed in Table 10.3.

It is worth mentioning that several controls can be used simultaneously, as shown in the next example, where both rp.doublebutton() and rp.slider() are applied to the same panel object. First, one defines the function which is called when an item is chosen, then one fills it with the values of the observed variable. Next, one draws the panel and places the proper function in it. The rp.control() function appears to be the central control function implemented in rpanel, since it is called

Table 10.3 Control functions

Function	Action
rp.doublebutton()	Adds a widget with "+" and "-" buttons
rp.checkbox()	Adds checkbox to the panel
rp.control()	Creates a control panel window
rp.listbox()	Adds a listbox to the panel
rp.radiogroup()	Adds a set of radiobuttons to the panel
rp.slider()	Adds a slider to the panel
rp.tkrplot()	Allows Rgraphics to be drawn in a panel

every time a new panel window is drawn, defining where the `rpanel` widgets can be placed. Eventually, `rp.slider()` and `rp.doublebutton()` are used in order to control a numeric variable by increasing or decreasing it with a slider or button widget.

The following code demonstrates the usage of both functions on dataset trees:

```
> require(rpanel)
> r = diff(range(Height))            # define the range of the variable
> density.draw = function(panel){   # draw density function
+    plot(density(panel$y, panel$sp))
+    panel
+ }
> # define panel window arguments
> density.panel = rp.control(title ="density estimation",
+    y  = Height,                     # data argument
+    sp = r / 8)                      # smoothing parameter
> # add a slider to the panel window
> rp.slider(density.panel, sp,
+    from   = r / 40, to = r / 2,    # lower and upper limits
+    action = density.draw,
+    main   ="Bandwidth")             # name of the widget
> # add a widget with"+"and"-"buttons
> rp.doublebutton(density.panel, sp,
+    step    = 0.03,
+    log     = TRUE,                  # step is multiplicative
+    range   = c(r / 50, NA),         # lower and upper limits
+    action= density.draw)
```

The first argument of `rp.slider()` identifies the panel object to which the slider should be added. The second argument gives the name of the component of the created panel object that is subsequently controlled by the slider. The `from` and `to` arguments define the start and end points of the range of the slider. The `action` argument gives the name of the function which will be called when the slider position is changed. The last argument adds a label to the slider (see Fig. 10.21).

The `rp.double.button()` function is used to change the value of the particular panel component by small steps when a more accurate adjustment of parameters is needed. Most of the arguments used by this function are the same as for the

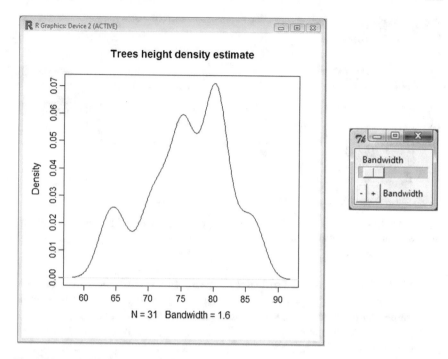

Fig. 10.21 Slider and double button for the control of density estimate.
Q BCS_ControlDensityEstimate

rp.slider(). The range argument serves the same purpose as the from and to arguments defining the limits for the variable.

Another feature enabled in rpanel is the possibility of interactively choosing between several types of plots to be applied to the same data set. It is also feasible to adjust different parameters within the chosen plot. This can be tested with rp.listbox(). This function adds a listbox of alternative commands to the panel. When an item is pressed, the corresponding graphics display will occur. The arguments of the function are the same as with the previous control functions. One can follow the setting of this function in the code below.

```
> data.plotfn = function(panel){          # define plot function
+   if(panel$plot.type =="histogram")      # choose histogram
+     hist(panel$y)                         # then plot histogram
+   else if (panel$plot.type =="boxplot")  # choose boxplot
+     boxplot(panel$y)                      # then plot boxplot
+   panel
+ }
> panel = rp.control(y = Height)           # new panel
> rp.listbox(panel, plot.type,             # list with 2 options
+   c("histogram","boxplot"),
+   action = data.plotfn,
+   title ="Plot type")                    # name of the widget
```

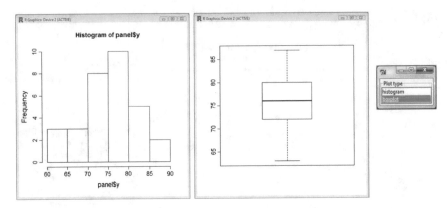

Fig. 10.22 Listbox control function with histogram and boxplot as alternative plots.
Q BCS_HistogramBoxplotOption

An alternative to `rp.listbox()` (see Fig. 10.22) is the function `rp.radiogroup()`, which can also be applied. The behaviour of this function is the same, but the panel has a set of radio buttons instead of a list view.

Another way to dynamically control the display output is offered by the function `rp.checkbox()`. This function adds a checkbox of alternative arguments for controlling the logical variables of the panel. When an item is selected, the corresponding graphics feature will be displayed.

The function `rp.tkrplot()` allows of placing the plot and widgets into one panel, creating one common window. `rp.tkrplot()` and `rp.tkrreplot()` call Tierney's `tkrplot` and `tkrreplot` functions, respectively, to allow Rgraphics to be displayed in a panel. This means that one should first define the plot of the function and then replot it inside the panel, as done below.

```
> if(interactive()){ # interactive navigation is allowed
+    draw = function(panel){
+       plot(density(panel$y, panel$sp),
+          col ="red", main ="")
+       panel
+    }
+    # place density plot and widget within one panel
+    redraw = function(panel){
+      rp.tkrreplot(panel, density)
+      panel
+    }
+    rpplot = rp.control(title ="Demonstration of rp.tkrplot",
+       y = Height, sp = r / 8)
+    rp.tkrplot(rpplot, density, draw)
+ }
```

The difference between this type of plot and the default plotting in `rpanel` can be observed in Fig. 10.23.

Fig. 10.23 Density plot with
`rp.tkrplot`.
Q BCS_rp.tkrplot

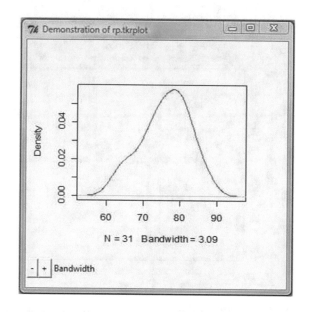

Table 10.4 Application functions

Function	Action
rp.ancova()	Plots a response variable against a covariate
rp.logistic()	Plots a binary response variable against a covariate
rp.plot3d()	Plots a 3D scatterplot, using the `rgl` package
rp.regression()	Plots a response variable against one or two covariates
rp.normal()	Plots a histogram and and fits the normal or other distributions to it

10.3.2 Application Functions in `rpanel`

The `rpanel` package also includes several useful built-in application functions.
These simplify the dynamic plotting of several processes, such as the analysis of
covariance, regression, plotting of 3D plots, fitting a normal distribution, etc. A list
of selected application functions is given in Table 10.4.

The `rp.regression()` function plots a response variable against one or two
covariates and automatically creates an `rpanel` with control widgets.

The arguments of the function are mostly relevant in the case of one covariate. So the use of `panel.plot` makes sense for two-dimensional plots and acti-
vates the `tkrplot` function in order to merge the control and output panels in
one window. One should be aware that three-dimensional graphics can not be
placed inside the panel. The code demonstrating the two-dimensional regression
with `rp.regression()` is presented below.

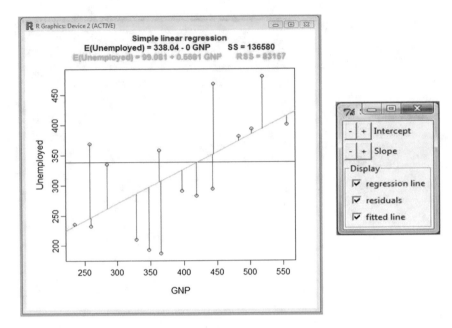

Fig. 10.24   Regression with one covariate. **Q** BCS_UnivariateRegression

```
> if(interactive()){          # interactive navigation is allowed
+    data(longley)
+    attach(longley)           # components are temporarily visible
+    # univariate regression
+    rp.regression(GNP, Unemployed,
+      line.showing = TRUE,  # regression line
+      panel.plot   = FALSE)# plot is outside the control panel
+ }
```

A regression line will appear in the plot if the argument line.showing is set to
TRUE. If the regression line is drawn, than one can interactively change its intercept
and slope, see Fig. 10.24.

   If the function has two covariates, the rp.regression() plot is generated with
the help of the rgl package, through the function rp.plot3d(), see Fig. 10.25.
In fact, one advanced interactive display will be created, which extends even the
features of the rgl 3D interactive scatterplot. The created plot is rotatable and
a zoom function is included. Additionally, one can set the panel argument to be
TRUE in order to create a control panel allowing interactive control of the fitted linear
models with one or two covariates. Double buttons are also available for stepwise
control of the rotation degrees of theta and phi.

```
> if(interactive()){   # interactive navigation is allowed
+    data(longley)
+    attach(longley)      # components are temporarily visible
+ # multivariate regression
+    rp.regression(cbind(GNP, Armed.Forces), Unemployed,
+    panel = TRUE)        # a panel is created
+ }
```

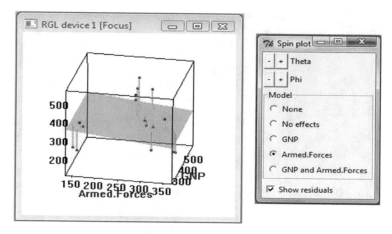

Fig. 10.25 Regression with two covariates. **Q** BCS_BivariateRegression

The rp.plot3d() function can be used independently from the rp.regression() function. With rp.plot3d(), a three-dimensional scatterplot would be created. The displayed plot would not differ from its counterpart created in the rgl package through the function plot3d().

The function rp.logistic() shares the same arguments and panel controls with rp.regression(), providing a basis for a logistic regression with a binary response variable.

Another application function is rp.ancova(), which provides the analysis of covariance, with different groups of data identified by colour and symbol. This function also shares its arguments with the function rp.regression().

```
> if(interactive()){     # interactive navigation is allowed
+    data(airquality)
+    attach(airquality)   # components are temporarily visible
+    rp.ancova(Solar.R, Ozone, Month,
+    panel.plot = FALSE)  # plot is outside the control panel
+ }
```

The function rp.normal() plots a histogram of data samples and allows a normal density curve to be added to the display. Furthermore, the fitted normal distribution with mean and standard deviation of the data sample can also be plotted. Double-buttons are built-in as well, and enable interactive control of the mean and standard deviation.

```
> if(interactive()){ # interactive navigation is allowed
+    y = Height        # data argument
+    # plot histogram with density curve
+    rp.normal(y, panel.plot = TRUE)
+ }
```

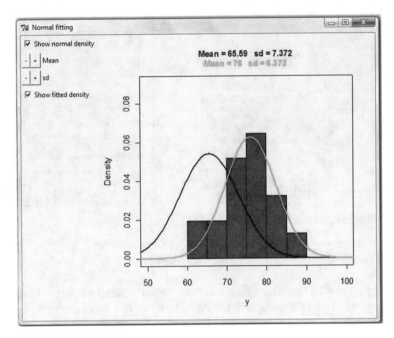

Fig. 10.26 Normal density fit. **Q** BCS_BivariateRegression

Figure 10.26 presents a screenshot of the observed function `rp.normal()`.

Bibliography

Adler, D., Nenadic, O., & Zucchini, W. (2003). RGL: A R-library for 3D visualization with OpenGL, *Technical report*, University of Goettingen.

Ahrens, J. H. (1972). Computer methods for sampling from the exponential and normal distributions. *Communications of the ACM, 15*(10), 873–882.

Ahrens, J. H., & Dieter, U. (1974). Computer methods for sampling from gamma, beta, poisson and binomial distribution. *Computing, 12*, 223–246.

Ahrens, J. H., & Dieter, U. (1982a). Computer generation of Poisson deviates from modified normal distributions. *ACM Transactions on Mathematical Software, 8*, 163–179.

Ahrens, J. H., & Dieter, U. (1982b). Generating gamma variates by a modified rejection technique. *Communications of the ACM, 25*, 47–54.

Albert, J. (2009). *Bayesian Computation with R, Use R!* (2nd ed.). New York: Springer.

Annamalai, C. (2010). Package "radx". https://github.com/quantumelixir/radx.

Ash, R. B. (2008). *Basic Probability Theory, Dover Books on Mathematics* (1st ed.). New York: Dover Pblications Inc.

Atkinson, A. C., & Pearce, M. (1976). The computer generation of beta, gamma and normal random variables. *Journal of the Royal Statistical Society, 139*, 431–461.

Babbie, E. (2013). *The Practice of Social Research*. Boston: Cengage Learning.

Banks, J. (1998). *Handbook of Simulation*. Norcross: Engineering and Management Press.

Becker, R. A., Cleveland, W. S., & Shyu, M.-J. (1996). The visual design and control of trellis display. *Journal of Computational and Graphical Statistics, 5*, 123–155.

Bolger, E. M., & Harkness, W. L. (1965). Characterizations of some distributions by conditional moments. *The Annals of Mathematical Statistics, 36*, 703–705.

Bowman, A., Crawford, E., Alexander, G., & Bowman, R. W. (2007). rpanel: Simple interactive controls for R functions using the tcltk package. *Journal of Statistical Software, 17*, 1–18.

Box, G. E. P., & Muller, M. E. (1958). A note on the generation of random normal deviates. *Annals of Mathematical Statistics, 29*, 610–611.

Braun, W., & Murdoch, D. (2007). *A First Course in Statistical Programming with R*. Cambridge: Cambridge University Press.

Brown, L. D., Cai, T. T., & DasGupta, A. (2001). Interval estimation for a binomial proportion. *Statistical Science, 16*, 101–117.

Broyden, C. G. (1970). The convergence of a class of double-rank minimization algorithms. *Journal of the Institute of Mathematics and Its Applications, 6*, 76–90.

Caillat, A.-L., Dutang, C., Larrieu, V., & NGuyen, T. (2008). Gumbel: package for Gumbel copula. *R package version 1. 01.*

© Springer International Publishing AG 2017
W.K. Härdle et al., *Basic Elements of Computational Statistics*,
Statistics and Computing, DOI 10.1007/978-3-319-55336-8

Canuto, C., & Tabacco, A. (2010). *Mathematical Analysis II. Universitext Series.* Milan: Springer.

Cheng, R. C. H. (1977). The generation of gamma variables with non-integral shape parameter. *Journal of the Royal Statistical Society, 26*(1), 71–75.

Cheng, R. C. H. (1978). Generating beta variates with nonintegral shape parameters. *Communications of the ACM, 21,* 317–322.

Clayton, D. G. (1978). A model for association in bivariate life tables and its application in epidemiological studies of familiar tendency in chronic disease incidence. *Biometrika, 65,* 141–151.

Cleveland, W. (1979). Robust locally weighted regression and smoothing scatterplots. *Journal of the American Statistical Association, 74,* 829–836.

Cook, R. D., & Weisberg, S. (1982). *Residuals and Influence in Regression.* New York: Chapman and Hall.

Cowpertwait, P. S., & Metcalfe, A. (2009). *Introductory Time Series with R.* New York: Springer.

Csorgo, S., & Farraway, J. (1996). The exact and assymptotic distributions of Cramér-von-Mises statistics. *Journal of the Royal Statistical Society Series B, 58,* 221–234.

Dalgaard, P. (2001). *The r-tcl/tk interface. Proceedings of DSC, 1,* 2.

Demarta, S., & McNeil, A. J. (2004). The t-copula and related copulas. *International Statiastical Review, 73*(1), 111–129.

Everitt, B. (2005). *An R and S-PLUS Companion to Multivariate Analysis.* London: Springer.

Everitt, B., & Hothorn, T. (2011). *An Introduction to Applied Multivariate Analysis with R.* New York: Springer.

Everitt, B., Landau, S., Leese, M., & Stahl, D. (2009). *Cluster Analysis.* Chichester: Wiley.

Fang, K. & Zhang, Y. (1990). *Generalized multivariate analysis,* Science Press and Springer.

Fishman, G. (1976). Sampling from the gamma distribution on a computer. *Communications of the ACM, 19*(7), 407–409.

Fletcher, R. (1970). A new approach to variable metric algorithms. *Computer Journal, 13,* 317–322.

Frank, M. J. (1979). On the simultaneous associativity of $f(x, y)$ and $x + y - f(x, y)$. *Aequationes Mathematicae. 19,* 194–226.

Frees, E., & Valdez, E. (1998). Understanding relationships using copulas. *North American Actuarial Journal, 2,* 1–125.

Gaetan, C., & Guyon, X. (2009). *Spatial Statistics and Modeling.* New York: Springer.

Genest, C., & Rivest, L.-P. (1989). A characterization of Gumbel family of extreme value distributions. *Statistics and Probability Letters, 8,* 207–211.

Genz, A. (1992). Numerical computation of multivariate normal probabilities. *Journal of Computational and Graphical Statistics, 1,* 141–150.

Genz, A. (1993). Comparison of methods for the computation of multivariate normal probabilities. *Computing Science and Statistics, 25,* 400–405.

Genz, A. & Azzalini, A. (2012). *mnormt: The multivariate normal and t distributions.* R package version 1.4-5. http://CRAN.R-project.org/package=mnormt.

Genz, A., & Bretz, F. (2009). *Computation of Multivariate Normal and t Probabilities.,* Lecture Notes in Statistics Heidelberg: Springer.

Genz, A., Bretz, F., Miwa, T., Mi, X., Leisch, F., Scheipl, F. & Hothorn, T. (2012). *mvtnorm: Multivariate Normal and t Distributions.* R package version 0.9-9993. http://CRAN.R-project.org/package=mvtnorm.

Goldfarb, D. (1970). A family of variable metric updates derived by variational means. *Mathematics of Computation, 24,* 23–26.

Gonzalez-Lopez, V. A. (2009). *fgac: Generalized Archimedean Copula.* R package version 0.6-1. http://CRAN.R-project.org/package=fgac.

Greene, W. (2003). *Econometric Analysis.* Upper Saddle River: Pearson Education.

Greub, W. (1975). *Linear Algebra. Graduate Texts in Mathematics.* New York: Springer.

Gumbel. E. J. (1960). Distributions des valeurs extrêmes en plusieurs dimensions. *Publications de Institut de Statistique de Université de Paris, 9,* 171–173.

Hahn, T. (2013). R2Cuba: Multidimensional numerical integration. http://cran.r-project.org/web/packages/R2Cuba/R2Cuba.pdf.

Härdle, W. K., & Vogt, A. (2014). Ladislaus von Bortkiewicz-statistician. *Economist and a European Intellectual, International Statistical Review, 83*(1), 17–35.

Härdle, W., Müller, M., Sperlich, S., & Werwatz, A. (2004). *Nonparametric and Semiparametric Models. Springer Series in Statistics*. New York: Springer.

Härdle, W., & Simar, L. (2015). *Applied Multivariate Statistical Analysis* (4th ed.). New York: Springer.

Hastie, T., Tibshirani, R., & Friedman, F. (2009). *The Elements of Statistical Learning: Data Mining, Inference, and Prediction*. New York: Springer.

Hestenes, M. R., & Stiefel, E. (1952). Methods of conjugate gradients for solving linear systems. *Journal of Research of the National Bureau of Standards, 49,* 409–436.

Hofert, M., & Maechler, M. (2011). Nested Archimedean copulas meet R: The nacopula package. *Journal of Statistical Software, 39*(9), 1–20.

Hoff, P. (2010). *sbgcop: Semiparametric Bayesian Gaussian copula estimation and imputation*. R package version 0.975. http://CRAN.R-project.org/package=sbgcop.

Ihaka, R., & Gentleman, R. (1996). R: A language for data analysis and graphics. *Journal of Computational and Graphical Statistics, 5*(3), 299–314.

Jarle Berntsen, T. E., & Genz, A. (1991). An adaptive algorithm for the approximate calculation of multiple integrals. *ACM Transactions on Mathematical Software, 17,* 437–451.

Jech, T. J. (2003). *Set Theory, Springer Monographs in Mathematics* (3rd ed.). The third millennium edition, revised and expanded: Springer-Verlag, Berlin.

Joe, H. (1997). *Multivariate Models and Dependence Concepts*. London: Chapman and Hall.

Joe, H., & Xu, J. J. (1996). *The estimation method of inference functions for margins for multivariate models, Technical Report 166*. Department of Statistics: University of British Columbia.

Johnson, M. E. (1987). *Multivariate Statistical Simulation*. New York: Wiley.

Johnson, P. (1972). *A History of Set Theory.*, Prindle, Weber & Schmidt Complementary Series in Mathematics Boston: Prindle, Weber & Schmidt.

Kachitvichyanukul, V., & Schmeiser, B. W. (1988). Binomial random variate generation. *Communications of the ACM, 31,* 216–222.

Kendall, M. G., & Smith, B. B. (1938). Randomness and random sampling numbers. *Journal of the Royal Statistical Society, 101*(1), 147–166.

Kiefer, J. (1953). Sequential minimax search for a maximum. *Proceedings of the American Mathematical Society, 4*(3), 502–506.

Knuth, D. E. (1969). *The Art of Computer Programming* (Vol. 2). Seminumerical Algorithms Reading: Addison-Wesley.

Kojadinovic, I., & Yan, J. (2010). Modeling multivariate distributions with continuous margins using the copula r package. *Journal of Statistical Software, 34*(9), 1–20.

Kruskal, J. (1964). Nonmetric multidimensional scaling: a numerical method. *Psychometrica, 29,* 115–129.

Kruskal, W. H., & Wallis, W. A. (1952). Use of ranks in one-criterion variance analysis. *Journal of the American Statistical Association, 47,* 583–621.

Marsaglia, G. (1964). Generating a variable from the tail of the normal distribution. *Technometrics, 6,* 101–102.

Marsaglia, G. (1968). Random numbers fall mainly in the planes. *Proceedings of the National Academy of Sciences of the United States of America, 61*(1), 25–28.

Marsaglia, G. (1995). Diehard Battery of Tests of Randomness, Florida State University.

Marsaglia, G., & Marsaglia, J. (2004). Evaluating the anderson-darling distribution. *Journal of Statistical Software, 9*(2), 1–5.

Marshall, A. W., & Olkin, J. (1988). Families of multivariate distributions. *Journal of the American Statistical Association, 83,* 834–841.

Martin, A. D., Quinn, K. M., & Park, J. H. (2011). Mcmcpack: Markov chain monte carlo in R. *Journal of Statistical Software, 42*(9), 1–21.

Matsumoto, M., & Nishimura, T. (1998). Mersenne twister: A 623-dimensionally equidistributed uniform pseudorandom number generator. *ACM Transaction on Modeling and Computer Simulations, 8,* 3–30.

McNeil, A. J. (2008). Sampling nested Archimedean copulas. *Journal Statistical Computation and Simulation, 78,* 567–581. (forthcoming).

Miwa, A., Hayter, J., & Kuriki, S. (2003). The evaluation of general non-centred orthant probabilities. *Journal of the Royal Statistical Society, 65,* 223–234.

Moore, E. (1920). On the reciprocal of the general algebraic matrix. *Bulletin of American Mathematical Society, 26,* 394–395.

Muenchen, R. A., & Hilbe, J. M. (2010). *R for Stata Users* (1st ed.). Statistics and Computing. New York: Springer.

Müller, H. (1987). Weighted local regression and kernel methods for nonparametric curve fitting. *Journal of the American Statistical Association, 82,* 231–238.

Nadaraya, E. (1964). On estimating regression. *Theory of Probability and Its Apllications, 9,* 141–142.

Nash, J. C. N., & Varadhan, R. (2011). Unifying optimization algorithms to aid software system users: optimx for r. *Journal of Statistical Software, 43*(9), 1–14.

Neave, H. (1973). On using the Box-Muller transformation with multiplicative congruential pseudo-random number generators. *Applied Statistics, 22,* 92–97.

Nelder, J. A., & Mead, R. (1965). A simplex method for function minimization. *Computer Journal, 7,* 308–313.

Nelsen, R. B. (2006). *An Introduction to Copulas.* New York: Springer.

Okhrin, O., Okhrin, Y., & Schmid, W. (2013). On the structure and estimation of hierarchical Archimedean copulas. *Journal of Econometrics, 173*(2), 189–204.

Okhrin, O. & Ristig, A. (2012). *HAC: Estimation, simulation and visualization of Hierarchical Archimedean Copulae (HAC).* R package version 0.2-5. http://CRAN.R-project.org/package=HAC.

Park, J. and Zatsiorsky, V. (2011). Multivariate statistical analysis of decathlon performance results in olympic athletes (1988-2008), *World Academy of Science, Engineering and Technology***77**.

Parzen, E. (1962). On the estimation of a probability density function and mode. *The Annals of Mathematical Statistics, 33,* 1065–1076.

Penrose, R. (1955). A generalized inverse for matrices. *Proceedings of the Cambridge Philosophical Society* (Vol. 51, pp. 406–413). Cambridge: Cambridge University Press.

Poisson, S.-D. (1837). *Probabilité des jugements en matière criminelle et en matière civile, précédées des règles générales du calcul des probabilitiés.* Paris: Bachelier.

Press, W. (1992). *Numerical Recipes in C: The Art of Scientific Computing.* Cambridge: Cambridge University Press.

Pyke, R. (1965). Spacings, Journal of the Royal Statistical Society. *Series B (Methodological), 27*(3), 395–449.

Quine, M. P., & Seneta, E. (1987). Bortkiewicz's data and the law of small numbers. *International Statistical Review, 55,* 173–181.

Development Core, R., & Team., (2012). *R: A Language and Environment for Statistical Computing.* Vienna, Austria: R Foundation for Statistical Computing. ISBN 3-900051-07-0. http://www.R-project.org/.

Razali, N. M., & Wah, Y. B. (2011). Power comparisons of Shapiro-Wilk Kolmogorov-Smirnov, Lilliefors and Anderson-Darling tests. *Journal of Statistical Modeling and Analytics, 2,* 21–33.

Rencher, A. (2002). *Methods of Multivariate Analysis.* New York: Wiley.

Richardson, L. F. (1911). The approximate arithmetical solution by finite differences of physical problems including differential equations, with an application to the stresses in a masonry dam. *Philosophical Transactions of the Royal Society A, 210,* 307–357.

Riedwyl, H. (1997). *Lineare Regression und Verwandtes.* Basel: Birkhaeuser.

Rosenblatt, M. (1956). Remarks on some nonparametric estimates of a density function. *The Annals of Mathematical Statistics, 27,* 832.

Samorodnitsky, G., & Taqqu, M. S. (1994). *Stable Non-Gaussian Random Processes*. New York: Chapman & Hall.

Sarkar, D. (2010). *Lattice: Multivariate Data Visualization with R*. New York: Springer.

Scherer, M., & Mai, J.-K. (2012). *Simulating Copulas: Stochastic Models, Sampling Algorithms, and Applications*. Series in Quantitative Finance. Singapore: World Scientific Pub Co Inc.

Serfling, R. J. (1980). *Approximation Theorems of Mathematical Statistics*. Wiley Series in Probability and Statistics. New York: Wiley.

Shanno, D. F. (1970). Conditioning of quasi-Newton methods for function minimization. *Mathematics of Computation, 24*, 647–656.

Shapiro, S. S., & Wilk, M. B. (1965). An analysis of variance test for normality (complete samples). *Biometrika, 52*, 591–611.

Shepard, R. (1962). The analysis of proximities: multidimensional scaling with unknown distance function. *Psychometrica, 27*, 125–139.

Sklar, A. (1959). Fonctions de repartition á n dimension et leurs marges. *Publications de Institut de Statistique de L' Université de Paris, 8*, 299–231.

Smirnov, N. (1939). On the estimation of the disrepancy between empirical curves of distribution for two independent samples. *Bulletin Mathématique de l'Université de Moscou, 2*, 2.

Stroud, A. H. (1971). *Approximation Caculation of Multiple Integrals*. New Jersey: Prentice Hall.

Theussl, S. (2013). Package "rglpk". http://cran.r-project.org/web/packages/Rglpk/Rglpk.pdf.

Trimborn, S., Okhrin, O., Zhang, S., & Zhou, M. Q. (2015). *gofCopula: Goodness-of-Fit Tests for Copulae*. R package version 0.2-5. http://CRAN.R-project.org/package=gofCopula.

van Dooren, P., & de Ridder, L. (1976). An adaptive algorithm for numerical integration over an n-dimensional cube. *Journal of Computational and Applied Mathematics, 2*, 207–217.

Venables, W. N., & Ripley, B. D. (1999). *Modern Applied Statistics with S-PLUS*. New York: Springer.

von Bortkewitsch, L. (1898). *Das Gesetz der kleinen Zahlen*, Leipzig.

Wasserman, L. (2004). *All of Statistics: A Concise Course in Statistical Inference*. New York: Springer.

Watson, G. (1964). Smooth regression analysis. Sankyah. *Ser. A, 26*, 359–372.

Weron, R. (2001). Levy-stable distributions revisited: Tail index >2 does not exclude the levy-stable regime. *International Journal of Modern Physics C, 12*, 209–223.

Whelan, N. (2004). Sampling from Archimedean copulas. *Quantitative Finance, 4*, 339–352.

Wilcoxon, F. (1945). Individual comparisons by ranking methods. *Biometrics Bulletin, 1*, 80–83.

Wuertz, D., many others and see the SOURCE file (2009a). *fCopulae: Rmetrics - Dependence Structures with Copulas*. R package version 2110.78. http://CRAN.R-project.org/package=fCopulae.

Wuertz, D., many others and see the SOURCE file (2009b). *fMultivar: Multivariate Market Analysis*. R package version 2100.76. http://CRAN.R-project.org/package=fMultivar.

Yan, J. (2007). Enjoy the joy of copulas: With a package copula. *Journal of Statistical Software, 21*(4), 1–21.

Index

© Springer International Publishing AG 2017
W.K. Härdle et al., *Basic Elements of Computational Statistics*,
Statistics and Computing, DOI 10.1007/978-3-319-55336-8

Printed in the United States
By Bookmasters